GENERAL CERTIFICATE OF EDUCATION

MODEL ANSWERS

BIOLOGY

(ORDINARY LEVEL)

Compiled by

G. Usher, B.Sc.

PUBLISHED AT

THE ARTEMIS PRESS

Sedgwick Park, Horsham, Sussex

General Editor: M. T. Bizony, M.A. (Cantab.)

First published October, 1962
Reprinted September, 1963
Reprinted November, 1969
Second Edition April, 1971
Reprinted June, 1972
Reprinted January, 1974
Third Edition July, 1975
Reprinted December, 1976

SBN 85141 285 8

Printed and bound in Great Britain by
REDWOOD BURN LIMITED
Trowbridge & Esher

THE SYLLABUS 1977 - '79

The main aim of most modern syllabuses is to encourage an understanding of basic principles. Facts should be used to illustrate principles and be clearly seen to do so. A grasp of the methods of science, of experimental procedures, of the unity which finds expression in the common physiological processes of plants and animals, and of the close links between form and function is greatly to be preferred to the memorizing of isolated facts.

The organisms studied should include:

a bacterial cell (outline structure as a simple unit of life); a simple protozoan; a filamentous green alga; a mould; Hydra or a sea anemone or another coelenterate polyp (feeding, reproduction, body plan, outline of life cycle); a moss or a fern (outline of life history); a tapeworm (adaptation to parasitic mode of life and its effect on health); a non-social insect, e.g. cockroach, housefly; a social insect, e.g. bee, ant (external characteristics common to all insects—the life-cycle of the non-social insect only. Organisation of activities within community of the social insect. Economic importance); a bony fish (emphasis on typical fish characters, adaptations to life in water); an amphibian (respiration, such development as can be seen with a hand lens); a herbaceous dicotyledon (externally visible structure of stem, root, leaves and flowers); a dicotyledonous tree (general shape and appearance throughout the year—details of flowers are not required); a mammal other than man (structural characters common to mammals, including internal features. Students should have the opportunity of seeing a dissected mammal).

Some syllabuses include a few additional forms. All of them should be viewed in the light of their relation to the following topics:

(1) Basic characteristics and the variety of living organisms.

(2) Nutrition, respiration and excretion.

(3) The importance of water; osmosis, diffusion, transpiration, and the internal transport of chemicals.

(4) Plant and animal behaviour; stimuli and responses; reflexes and tropisms; nervous and chemical mechanisms; movement.

(5) Reproduction, development and growth.

(6) Life and the environment; ecological considerations; adaptation; the parasitic mode of life; human activity and its effects; conservation.

CONTENTS

	PAGE
General Advice	**5**
Mixed Questions	**7**
Morphology and Anatomy	**9**
Ecology	**19**
Life Histories	**23**
Plant Physiology	**33**
Animal Physiology	**47**
Nutrition	**54**
Reproduction	**59**
Irritability	**71**
Circulation	**73**
Multiple Choice Questions	**74**
Model Test Paper Index	**80**
Answers to Multiple Choice Questions	**80**

The Publishers are indebted to the University of London for permission to reproduce questions from the G.C.E. Examination.

GENERAL ADVICE

Obey the instructions given at the head of the examination paper absolutely. Compulsory questions may carry more marks than the others so be sure that they are done well. Divide your time equally between the questions; if you spend too long on one question, you are jeopardizing your chances of gaining marks on other questions. Each question carries a maximum number of marks which cannot be exceeded, however good your answer may be.

DIAGRAMS should be large and clear, drawn in pencil, and labelled in ink. The indicating lines should touch the structures to which they refer, and the lines of the drawing should be thin and continuous — not sketched. Avoid the use of coloured pencils, and reduce shading to a minimum; they waste time and often lead to confusion rather than clarification.

Many marks can be lost by careless reading of the question. Here are a few examples:—

Q: '*Describe briefly . . .*'

Such a question usually has several parts, so that each part can be given only a short time. Brevity is essential.

Q: '*Using a named example . . .*'

Ensure that the example is named, and that the subsequent description applies to this example.

Q: '*Give three examples . . .*'

An equal number of marks will be allocated to each example. Do not leave any out, or the marks for that part of the question will automatically be lost.

Q: '*Give an illustrated account . . .*'

A definite number of marks will be allocated to the illustrations. This type of question can often be answered by well-labelled drawings. Writing can be reduced to a minimum, and be used only to explain points which cannot be made clear in the diagrams.

Avoid irrelevancies. For example, if you are asked to describe the structure of an organ, there is no need to go into great detail as to its function.

A question asking for a comparison cannot be satisfactorily answered simply by describing the things to be compared one after the other. A good comparative essay stresses both the points of similarity and the points of contrast throughout (see QQ. 4 & 20). Occasionally such a question can be answered by a careful tabulation, as in Q. 27.

Some Examining Boards set a short answer paper. Although it is short, it is very searching; candidates frequently treat it too lightly and do not give precise answers to the questions.

A PRACTICAL EXAMINATION is often set. Do exactly what you are told on the paper, *e.g.* do not make drawings if only a list is required. Know the differences between dorsal, lateral and ventral views, and between longitudinal and transverse sections.

DRAWINGS should be as large as the paper permits, and as detailed as possible. Pay attention to the proportions of your drawings, and arrange the specimen to show as many features as you can within the limits of the question. Label all the parts, however obvious or irrelevant such labelling may seem, and cut down shading to a minimum.

Good English and orderly production of material are essential. Divide your answer into parts in the same way in which the question is divided. Practice a clear, unambiguous style, and avoid 'padding'.

Mixed Questions

Q. 1:

Answer briefly any FOUR of the following:—

(*a*) Why do many plants wilt on a hot, windy day?

(*b*) What are the functions of blood cells (corpuscles)?

(*c*) What are the functions of the skin of a mammal?

(*d*) Why does a man lose his balance after turning round quickly several times?

(*e*) Why are potatoes grown from tubers?

[*Although the question asks for brief answers, about seven minutes are available for each part, and in that time you should be able to state concisely all the relevant major points.*]

A: (*a*) Wilting is caused by the rapid and excessive loss of water from a plant. When the uptake of water by the plant roots is not fast enough to make good this loss the cells of unlignified tissue become flaccid. The loss of turgor causes the cell walls to collapse, and consequently the plant wilts.

An increase in the atmospheric temperature lowers the relative humidity and so increases the difference between the concentration of water vapour in the atmosphere and in the intercellular spaces of the plant. This in turn increases the rate of diffusion of water vapour from the plant to the atmosphere through the stomata and lenticels. If the atmosphere is still, the water vapour diffusing out from a pore in the plant tends to remain in the immediate vicinity of the pore, raising the relative humidity of the micro-climate and reducing evaporation. If the atmosphere is disturbed, as on a windy day, this water vapour is removed and replaced by relatively dry air, which increases the rate of diffusion from the plant.

(*b*) The red blood cells are anucleate biconcave discs produced in the bone marrow, and contain the red haemoglobin. They act as vehicles for the haemoglobin which combines with oxygen so that the oxygen is carried to the body tissues. In the lungs, where the oxygen concentration is high, the haemoglobin forms a loose chemical combination with it to form oxyhaemoglobin which is bright red. The blood is then carried to the tissues; where there is an oxygen deficit, the oxyhaemoglobin gives up its oxygen which is used by the cells. The haemoglobin is then in its stable state, and is bluish-red.

The white blood cells are nucleate amoeboid cells whose function is to protect the body against bacterial invasion. They put out pseudopodia and engulf the bacteria and digest them. They also produce antitoxins which neutralise toxins (poisons) secreted by the bacteria, and destroy injured or infected tissue.

The blood platelets are anucleate and about a quarter of the size of the red blood cells. They assist in the clotting of blood.

(*c*) The functions of the skin are as follows:—

(i) Protection. The horny layer of the epidermis, produced by the cells below it, affords protection to the underlying structures. This layer is dead and fairly tough, reducing the risk of mechanical and chemical injury. It prevents invasion by bacteria, and stops the indiscriminate loss of water by evaporation from the body surface.

(ii) Control of body temperature. The fat in the dermis acts as an insulator against excessive heat loss, and under cold conditions the erector muscles contract, lifting the hairs on the surface. This causes a layer of still air to be trapped against the surface of the skin, and still air is a very good insulator. If the body temperature increases above normal, the sweat glands produce perspiration on the surface of the skin. This is evaporated by the heat of the body, causing a loss of heat and a reduction in body temperature.

(iii) Excretion. Sweat contains salts, water, carbon dioxide and urea which are excretory products. The excretory function of the skin is relatively unimportant.

(iv) Sensory functions. The skin contains sensory cells and nerve endings which are sensitive to heat, cold, pain and pressure.

(v) Storage. Fat is stored in the dermis.

(vi) Vitamin D is synthesised in the presence of ultra-violet light in the skin of man.

(*d*) The organs of balance in man are the utriculus and semicircular canals of the inner ear. These, with other organs, make up the labyrinth which floats in the perilymph. The labyrinth is also filled with fluid called the endolymph. The three semicircular canals are mutually at right angles, and attached to the utriculus. At the distal end of each canal is a swollen ampulla containing chalky particles and sensory hairs. The arrangement of the canals is such that any movement of the head must cause the movement of the endolymph in at least one of the canals. This agitates the chalky particles, knocking them against the sensory hairs. The sensation is transmitted through the auditory nerve to the cerebellum which sends an impulse to the appropriate muscles to control the movement and restore balance. If the head is turned quickly several times, the endolymph will continue to move under its own inertia after the head has stopped moving. This means that the cerebellum will continue to send out impulses which will cause the muscles to correct a movement which has in fact stopped, resulting in the man losing his balance.

Morphology and Anatomy

Q. 2:

Make a labelled drawing of the root system of a NAMED annual plant and a further labelled drawing on a larger scale to show the tip of one root. Make a labelled drawing of the shoot system of the same or a different NAMED plant. Use these two drawings to indicate the essential differences between shoot and root.

A:

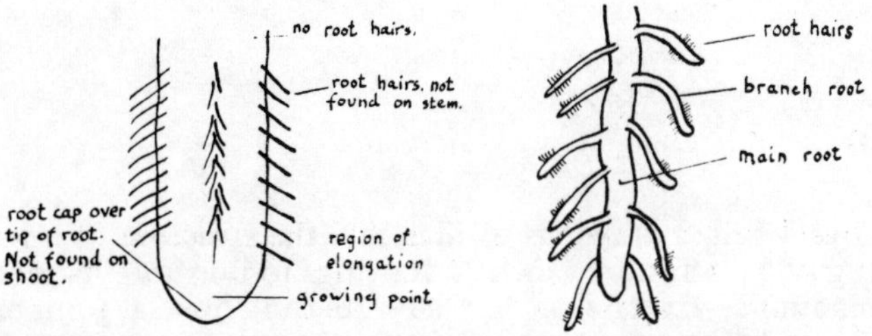

BROAD BEAN: Root Tip and Root System

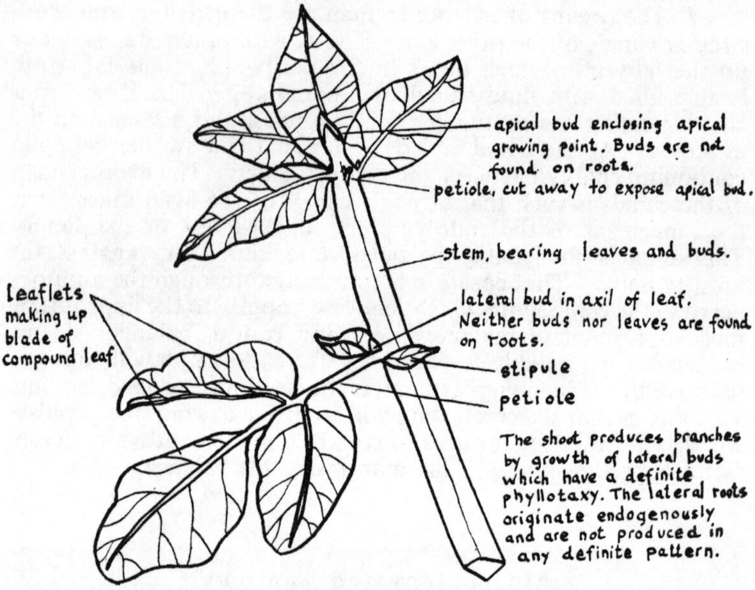

BROAD BEAN: Part of Shoot

[*If the drawings are sufficiently well labelled, there is no need for any written part for this question.*]

Q. 3:

Give labelled drawings to illustrate the structure of a NAMED seed: give TWO further labelled drawings to illustrate its mode of germination. Explain, as far as you can, how a germinating seedling obtains energy until it is able to carry out photosynthesis.

A:

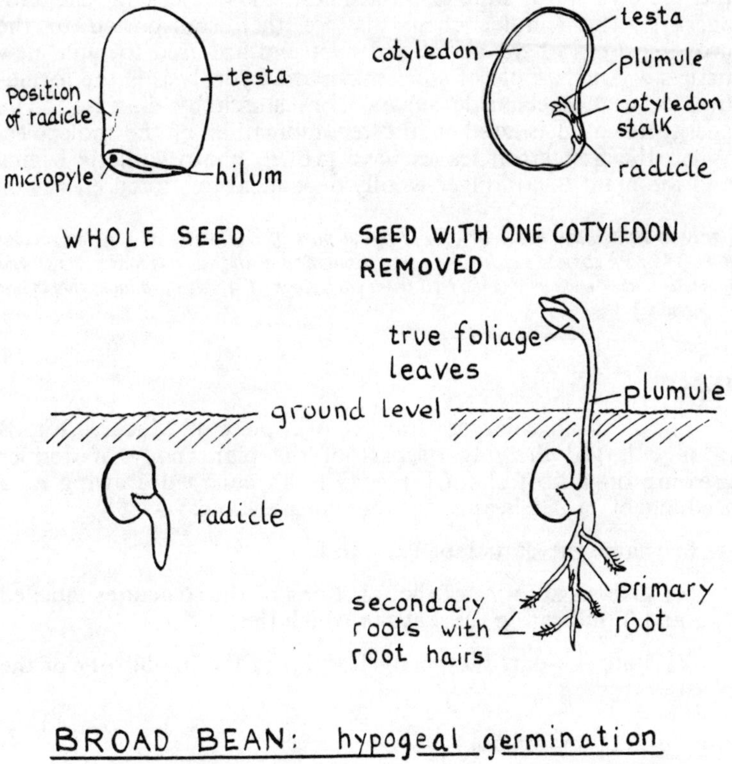

The energy needed by the germinating seedling is stored as chemical energy in the food reserves of the seed. These reserves usually consist of foods with a high calorific value, *e.g.* the oil in the endosperm of the Castor Oil seed, the protein in the cotyledons of the Bean seed, and starch in the endosperm of the Rice seed.

When a seed is dormant, it has a low water content, so that respiration and the other life processes are proceeding slowly. In the initial stages of germination, water is rapidly absorbed through the testa, enabling foods to be dissolved, so that the life processes can be quickened. Enzymes are secreted by the part of the plant in contact with the food, *e.g.* the hypocotyl in the

Broad Bean. These enzymes digest the food reserves, breaking them down into simpler molecules, thus releasing chemical energy. The simpler chemicals are then transported to the growing areas of the shoot and root and are used to build new tissues. An example of such enzymatic hydrolysis is the formation of monosaccharide sugars from starch by diastase. The energy released is used in the recombination of the molecules. When the first green leaves have grown, photosynthesis begins and the plant is no longer wholly dependent on stored energy.

[*Only drawings are necessary for the first part of the question. In the second part it is very easy to get involved in a discussion of the various materials stored in seeds and how energy is derived from each form of storage product; this is not necessary.*]

Q. 4:

Fig. A is an enlarged drawing of a piece of Mucor. Fig. B is an enlarged drawing of part of the plant body of dodder growing on a host shoot. Fig. C is an enlarged drawing of a seedling of broad bean. In your answer-book:—

(*a*) name the parts labelled 1 to 8,

(*b*) give an account of the functions of the structures labelled 2, 5 and 8, indicating the ways in which they differ,

(*c*) state the part played by 1 and 3 in the life-history of the plants.

A: (*a*) (1) sporangium
(2) hypha
(3) flower head of dodder
(4) dodder stem
(5) haustoria of dodder
(6) plumule bearing true leaves
(7) cotyledon enclosed by testa
(8) radicle with branch roots

(*b*) The hyphae of Mucor, the haustoria of dodder, and the roots of the Broad Bean are absorbing organs, taking essential materials into the organism. The hyphae of Mucor are embedded in the substrate on which it is saprophytic, and the hyphal tips

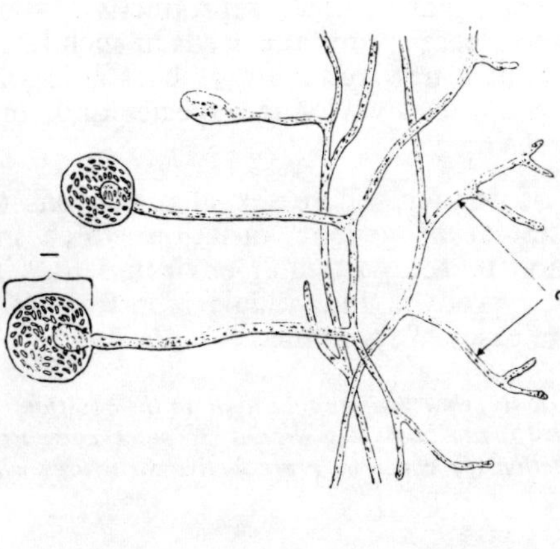

Figure A

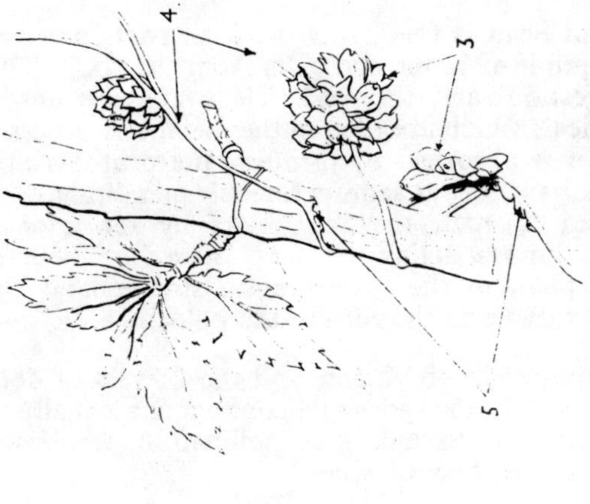

Figure B

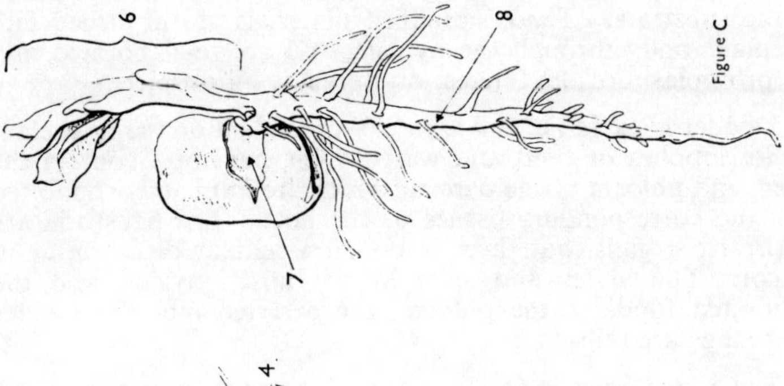

Figure C

produce enzymes which bring about the extra-cellular digestion of the substrate. These simplified materials are absorbed into the plant body through the hyphae, and are re-elaborated into the protoplasm of the fungus, or stored as oil droplets.

Dodder is parasitic, and is totally dependent on the host plant for its supplies of food and water. The haustoria contain the xylem and phloem of the parasite which are intimately connected with the corresponding tissues of the host. The haustoria are absorptive organs, but there is no extra-cellular digestion as in Mucor. The water and salts in the host's xylem, and the elaborated foods in the phloem, are diverted into the dodder where they are utilized.

The broad bean is free-living, and the roots are used in the active absorption of water and salts from the soil. There is no form of digestion, and the materials which are absorbed are simple chemicals which are used in the metabolic processes of the plant. Water is absorbed by osmosis, the protoplasmic linings of the root hairs acting as semi-permeable membranes. Salts are also absorbed through the root hairs by ion exchange, the protoplasmic linings acting as ionic 'sieves'. These materials pass up the plant in the transpiration stream, and elaborated materials pass down to the root in the phloem.

(*c*) The sporangia of Mucor, and the flowers of dodder, are reproductive organs, the sporangia containing asexually produced spores, and the flowers aiding in pollination, fertilization, and the ultimate production of seeds.

In both these plants, the reproductive structures are abundant. They release spores and seeds in such large quantities that colonization of a new substrate or host is highly probable, thereby ensuring the survival of the species and, in favourable conditions, increasing it.

In the case of dodder, where sexual fusion has taken place, there is a possibility of genetic variation which may lead to further adaptation to the particular environment. The asexual spores of Mucor produce individuals identical to the parent, except in the rare cases of spontaneous mutation.

[*Be careful to do exactly what you are told to do in the question. All the parts indicated in section* (*b*) *are absorbing organs, so some comparison should be given: similarly section* (*c*) *concerns reproductive structures which should be compared.*]

Q. 5:

Describe the kidneys, ureters and the associated blood vessels in a mammal. Give a large labelled diagram to show the position of these organs in the body and how they are connected with the exterior. State the functions of the kidney.

[*This question necessitates the description of the internal structure of the kidney. The diagram must include the bladder and urethra, as the question covers connection of the kidney to the exterior.*]

A:

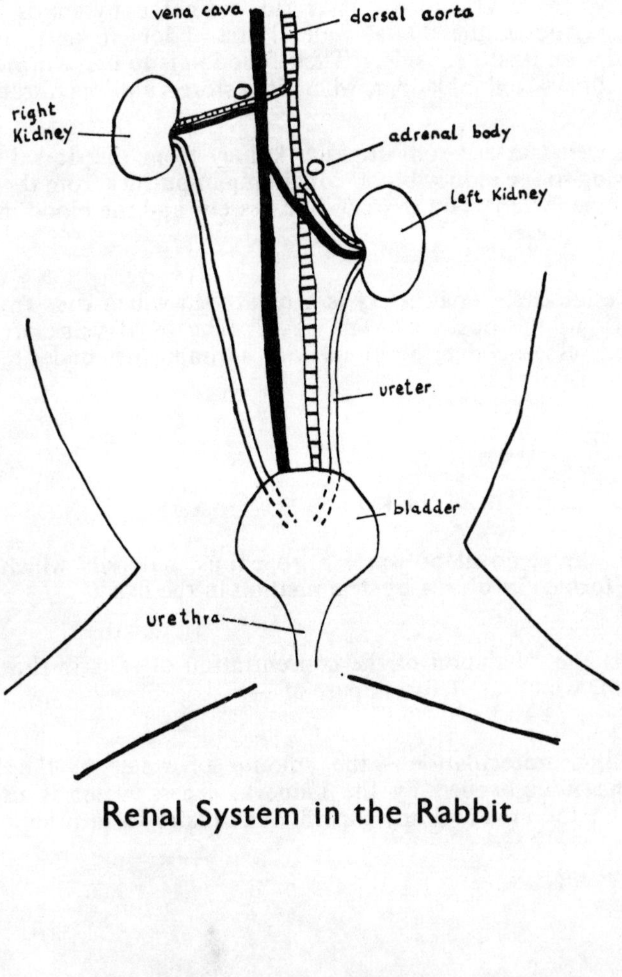

Renal System in the Rabbit

The kidneys are bean-shaped organs situated high in the abdomen on the dorsal wall. They are dark red, and the left kidney is lower than the right. In longitudinal section the kidney is clearly differentiated into external cortex and internal medulla, which converges into the pyramid. The pyramid then opens into the ureter. The cortex consists of a mass of Malpighian bodies and their tubules, intertwined with capillaries between branches of the renal artery and vein, while the medulla is made up of the uriniferous tubules which converge on the pyramid.

The ureters arc relatively thin-walled tubes running from the kidneys to the bladder. They run from the pyramids of the kidneys, down the dorsal wall of the abdomen, and into the bladder on its dorsal side. Their function is to carry urine from the kidneys to the bladder, where it is stored until micturation.

A renal artery runs to each kidney from the dorsal aorta, carrying to the kidney blood containing impurities from the body. The renal arteries are thick-walled vessels, and the blood in them is under pressure.

Beside each renal artery is a renal vein which runs from the kidney to the posterior vena cava. The renal veins are thin-walled vessels conveying blood without impurities under reduced pressure.

The functions of the kidneys are:—

(i) the removal of waste nitrogenous materials which have been formed into urea by deamination in the liver;

(ii) the regulation of the concentration of salts in the blood stream, which itself forms part of —

(iii) osmoregulation — the amount of water in the blood plasma is controlled by the kidneys; excess water is used to dissolve the urea and any superfluous salts to form urine.

Q. 6:

Make a large labelled drawing to show the position and attachment of the vertebrate eye to its orbit; namcs of the eye muscles are not required. Make a further labelled diagram of a vertical section of a vertebrate eye. Explain concisely how a change of focus is brought about.

A:

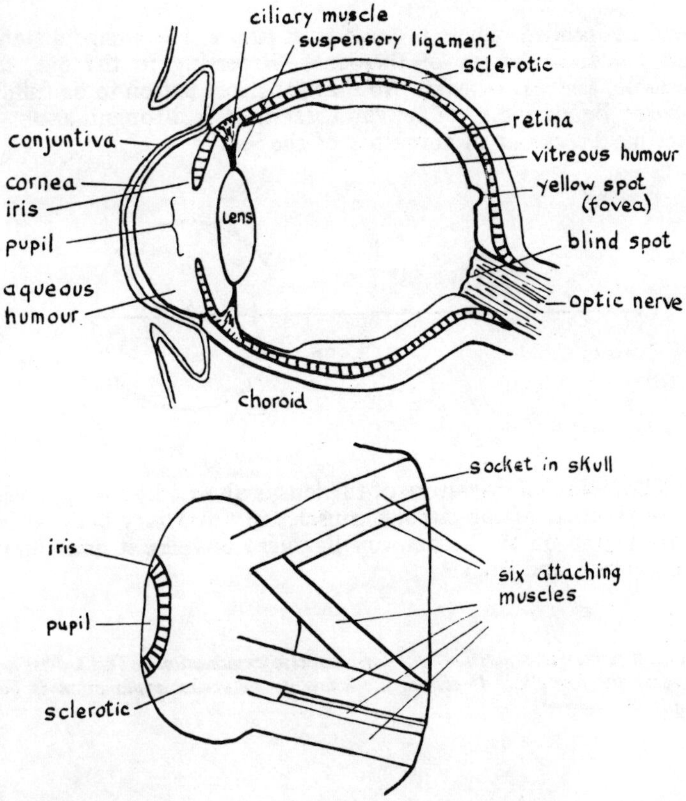

L.S. and attachment of VERTEBRATE EYE

The cornea, aqueous humour, lens, and vitreous humour are all transparent, and refract light to form a clear image of an external object on the retina. Only the thickness of the lens can be altered to focus an image.

For an image to be in focus, the condition

$$\frac{1}{u} + \frac{1}{v} = \frac{1}{f}$$

must be fulfilled, where u = object distance, v = image distance, and f = focal length of the lens. Referring to the eye, u is variable, and v is constant, so that for the equation to be fulfilled f must be variable. The variation in f is brought about by altering the radius of curvature of the lens.

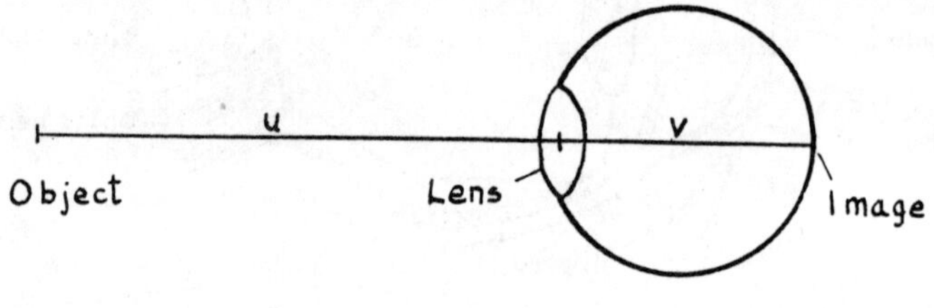

The radius of curvature of the lens is altered by the expansion or contraction of the circular muscles in the ciliary body which, in turn, pull on the suspensory ligament or relax it, making the lens thinner and thicker.

[*The last part of the question asks for a concise explanation. This is best done by using the formula. If you do not know it, the verbal explanation is quite simple, but longer.*]

Ecology

Q. 7:

What is soil? Describe the ways in which the chemical composition and physical structure of soil is affected by the activities of plants and animals.

[*This is a very wide question, and has to be treated carefully. The first part is straightforward, but the second part needs reference to a wide range of plants and animals. It is advisable in such a question to deal briefly and concisely with as many different aspects of the subject as possible, and not to dwell too long on one topic.*]

A: Soil is the product of the physical and chemical weathering of rocks. When it is formed and remains on top of the parent rock it is called *sedentary*; if it has been deposited by rivers, glaciers or the sea it is called *sedimentary* or *alluvial*. Chalk soils and coral sands consist almost wholly of the fossil remains of marine organisms with calcareous shells.

Soil is made up of minerals, air and water, and the top-soil contains in addition living organisms and organic débris or humus. Some of the minerals are sparingly soluble, and are utilized by plants growing in the soil, but the bulk of the soil substances are insoluble. If the soil particles are generally small, and colloids are formed, the soil is a clay, and if they are relatively large, the soil is sandy. Clay soils hold water and have little air in them, while sandy soils are free-draining and well-aerated. The lack of water gives a sand a low specific heat, so that it responds rapidly to changes in atmospheric temperature. The opposite is true of clays.

The humus is made up of decayed organic material. It is responsible to a large degree for the brown colour of the soil. Humus supplies plants with minerals which were once part of the organism from which the humus was derived. It is of great value in soil in that it holds water in quick-drying sands, and loosens the texture of heavy soils by its fibrous constituents.

The soil water is a dilute solution of carbon dioxide (carbonic acid) and has soluble minerals dissolved in it. If the soil is water-logged, the spaces between the particles are filled with water, but under normal conditions the water is held on the soil particles by capillary attraction, and in the soil colloids by imbibition. It is only the capillary water that is available to plants.

The air in the soil is essential for the respiration of soil organisms and plant roots, and is contained in the spaces between the particles. The nitrogen in the air is utilized by the nitrogen fixing bacteria.

The mineral constituents of the soil may be altered by animals. For example, molluscs use calcium in making their shells, and when they die, the zone around the shell will have a high concentration of calcium which it did not have before. Calcium and phosphorus are extracted from the soil by plants, which may later be eaten by animals. The elements are then taken into bone tissue, which will form local concentrations of calcium and phosphorus in the soil when the animals die.

Bacteria and fungi which cause decay release soluble substances into the soil. A specific instance of this is the formation of nitrates from ammonia by bacteria in the nitrogen cycle. The concentration of nitrogen-containing substances in the soil is also increased by the nitrogen-fixing bacteria found in the root nodules of legumes and other plants. The soil immediately around the roots of plants (the *rhizosphere*) contains compounds exuded by the roots that are not found in barren soil.

Humus is added to the soil by decaying plant material, the dead bodies of animals, and the faeces of animals. The level of concentration and the condition of the humus depend on the presence of the decay-bacteria and fungi. For them to thrive, there must be sufficient air in the soil. If the soil is water-logged, the organic matter remains undecayed as raw humus, and an extreme example of this is a peat bog. If the humus is well-decayed, a Brown Earth soil is formed. This has plenty of minerals for plant growth, is well-aerated and holds sufficient water.

Animals such as earthworms and insect larvae burrow in the soil, improving the drainage and allowing the air to penetrate. Large herbivorous animals living on the surface of the soil can affect its structure by denuding it of vegetation, and consequently causing soil erosion. Dead plant roots decaying in the soil leave humus filled channels which aid aeration.

Sometimes a particular type of vegetation and its associated animal life affect the soi in a way which gradually renders it less favourable for this typeland more favourable for another type of vegetation. This is an important factor in ecology, and can lead to a succession of different types of flora, in some cases within a few years, in others over a period of centuries.

Q. 8:

Describe the methods of dispersal of fruits and seeds. Illustrate your answer with labelled drawings of THREE NAMED fruits or seeds each suited for a different method of dispersal.

A: Dispersal of seeds (or of the fruits containing them) is necessary for the survival of plant species. It prevents overcrowding of a particular area by one species, and gives an opportunity for the colonization of new areas.

Dispersal is brought about in four ways: (i) by animals, (ii) by wind, (iii) by water, and (iv) by special explosive mechanisms.

The carrying of seeds by animals may be internal or external. In the former case, the fruits are attractive to animals, and are eaten, the seeds being passed out undamaged in the faeces, *e.g.* blackberry or tomato. External transport may take many forms: the adhesion of the sticky seeds of mistletoe to the bills of thrushes; nuts being transported by rodents and then forgotten; fruits being eaten and the stones discarded; or fruits and seeds being carried in the mud on the feet of animals, or on the clothing of human beings. Man plays an important part in the inter-territorial dispersal of seeds in commercial products, *e.g* in wool. Some plants (goosegrass, avens, *etc.*) rely on hooks to facilitate transport of their seeds. These hooks catch in the fur of animals and the fruit may be transported some distance before it is rubbed off.

Wind dispersal takes three main forms. The seeds of some plants (*e.g.* orchids) are so light and small as to be blown as dust particles by wind. Other plants (*e.g.* the Poppy) have a censer mechanism; the dry seeds are free in the fruit, from which they are shaken through pores when the long-stalked fruit is agitated by the wind. The third method depends on increasing the surface area of a relatively large seed so that it can be carried by the wind. The sailed fruits of sycamore, the winged seeds of pine, and the parachute mechanisms of the fruits of dandelion are all adaptations of this principle.

Decreasing of the density of water-dispersed seeds by a large bulk of light material enables the seeds to float, *e.g.* the mesocarp of the coconut is fibrous and can drift over long distances between the islands of the Pacific Ocean.

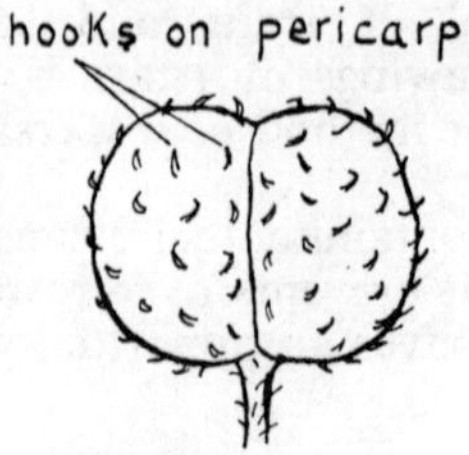

GOOSEGRASS FRUIT

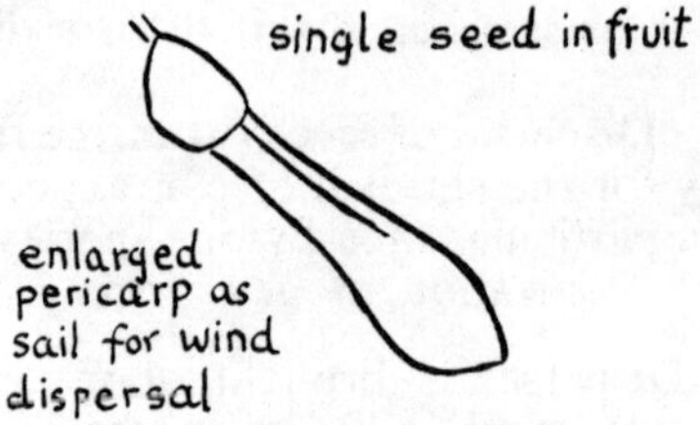

SYCAMORE FRUIT

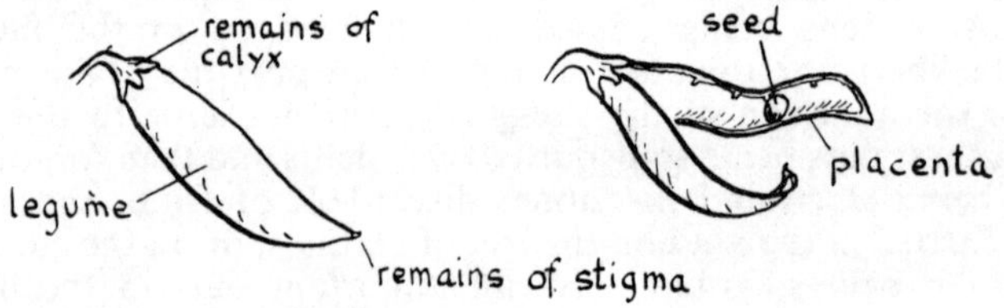

PEA POD. before and after dehiscence

Explosive mechanisms depend on the developing of tensions (usually by differential drying) within the fruit. When these tensions reach a maximum, the fruit bursts suddenly, scattering the seeds. Such mechanisms are found in the legumes, and in the violet.

Life History

Q. 9:

With reference to plants useful to man, explain what you understand by the following:

(*a*) an annual, (*b*) a biennial, (*c*) a perennial herb, (*d*) a tree. Give a short description of the part of the plant, in each of your examples, which is used by man. (Give only ONE example in each category.)

A: (*a*) An annual is a plant which completes its life-cycle in one year. The plant dies at the end of the season, and the species survives the winter as seeds in the soil, which germinate later to produce new plants. The flowers develop and, after pollination and fertilization, seeds are formed. The parent plant dies after seed-dispersal.

An example is the Broad Bean. The parts of the plant useful to man are the seeds, which have a high protein content and are eaten as a vegetable. The seeds are enclosed in the fruit, which in this case is a pod, or *legume*.

(*b*) A biennial plant needs two years to complete its life-cycle. The seeds germinate in the spring of one year, and the plant builds up a food store in specialized storage organs during the first year. The aerial parts die down in the winter, and the plant survives as an underground storage organ. In the second spring and summer, foliage leaves are produced, as are the flowers later in the season. The seeds are formed and dispersed, and the parent plant then dies.

An example of a biennial is the Sugar Beet, which is useful as the storage product is sucrose. Beet is cropped as an annual, as only the storage organ is needed for consumption. The parts of the Beet used for the production of sugar are the stem, hypocotyl and root. The leaves of Sugar Beet are sometimes fed to stock, or made into silage. Occasionally, seeds are produced in the first year. Plants behaving in this way are called 'bolters'.

(*c*) A perennial herb survives for more than two years, usually producing flowers and seeds after the first year. The green, aerial parts shoot annually from an underground storage organ, and die down each winter. The underground parts are frequently

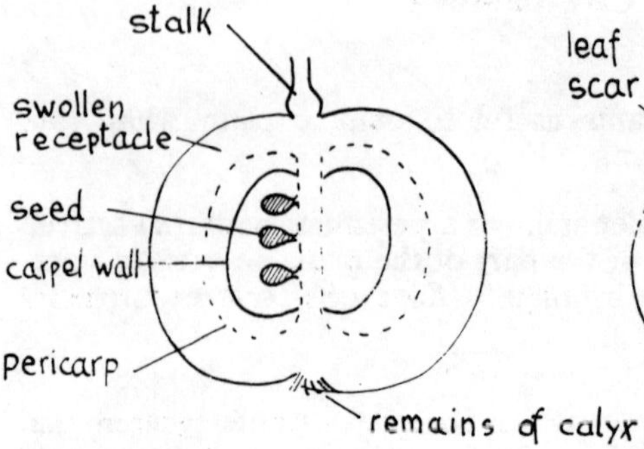

L.S. APPLE

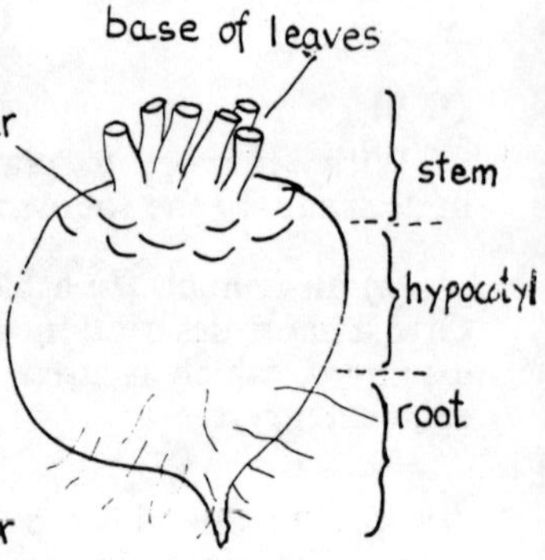

SUGAR BEET

capable of vegetative reproduction. The Onion is an example of a perennial herb useful to man.

The Onion is a bulb, consisting of a compressed stem, bearing leaves and adventitious roots. The leaves are of three types. The external, papery scale leaves protect the fleshy leaves which make up the bulk of the bulb, and which store sugars and water. The apical growing point of the shoot produces a stem bearing leaves and flowers. These elongate, and grow up through the central space between the fleshy leaves. An axillary bud of a storage leaf swells up during the growing season to form a new bulb which survives the winter.

(*d*) A tree is a perennial plant which shows a great deal of secondary thickening, leading to the production of wood. It usually has a single trunk, and lives for much longer than three years. Trees are distinguished from shrubs by their greater size.

The Apple Tree is used by man, the fruits being eaten. This fruit is a source of sugar and vitamins.

The apple is a false fruit, being made up of a swollen receptacle, as well as a swollen ovary. This type of fruit is called a *pome*.

[*In spite of the way in which the question is worded, it is probably easier to describe the examples under the sections, rather than all together in a final section.*]

Q. 10:

Illustrate seasonal change by reference to the lives of TWO animals and TWO plants.

[*Define 'seasonal change' and then select your examples to illustrate the different points as far as possible.*]

A: Seasonal changes are temporary alterations in the physiology or morphology of an organism in response to climatic changes in environment, They include such phenomena as hibernation, encystment, changes of the colour of fur or plumage, and the physiological changes linked to breeding seasons and migration, and form part of the reproductive cycle in most flowering plants.

THE BAT

The Bat provides a good example of hibernation. Bats are colonial mammals which feed on insects they catch in flight during the summer nights. The population of adult insects is greatly reduced during the winter months, and bats do not migrate to any great extent. They therefore feed voraciously during the summer and store the excess food as a layer of fat under the skin as a reserve for the winter. Hibernation is brought about probably as much by the accumulation of fat as by the seasonal drop in temperature. In hibernation the animal is in a state of torpor: the body temperature drops almost to the temperature of the surroundings, and the heart, breathing and metabolic rates are greatly reduced. The stored food is therefore consumed very slowly and is usually sufficient to last until the warmer weather returns and the bats become active again.

THE HOUSE-FLY

Like any insect with a complete metamorphosis, the fly shows seasonal variation in the changes from egg to larva, from larva to pupa, and from pupa to adult. The change from larva to pupa is the crucial one, and this is brought about by changes in the physiology of the insect, *i.e.*, by changes in the internal environment. Changes take place in glands at the base of the brain which influence growth and moulting of the larva, and the ultimate formation of the pupa.

The activity of flies is influenced by the external temperature, and when this drops the flies become torpid, and it is in this condition that a few adult flies survive the winter. This is not hibernation in the strict sense of the word.

Horse-Chestnut

During the winter the tree is dormant, and devoid of leaves. With the warmer weather of spring and an increase in soil temperature, the sap begins to rise, and the dormant buds burst to produce green leaves and flowers. The flowers are pollinated and the fruits formed, and the tree remains vegetative for the rest of the season. When the soil becomes cooler, the temperature is not sufficiently high for the roots to take in salts and water efficiently. This condition is called physiological drought. Because of the lack of water there is a need for the reduction of water loss by transpiration. This reduction is effected by leaf fall. A layer of cork, called the abscission layer, is formed across the base of the petiole of each leaf, cutting off its supply of water. The leaf gradually turns brown and dies, ultimately falling off the tree.

Seasonal change is also illustrated by the growth of the Horse-Chestnut tree. Branches increase in length by the extension growth of the vegetative terminal buds, and in thickness by secondary thickening, caused by the activity of the cambium.

Daffodil

The daffodil bulb is made up of fleshy leaves attached to a flattened stem with an apical bud. During the winter the bulb lies dormant in the soil, and there is little biological activity. In the spring the apical bud extends rapidly, and grows up through the centre of the bulb to produce green foliage leaves and the flower stalk. The flowers are pollinated, and seeds are produced. The flower then dies and the plant remains vegetative for the rest of the season. The green leaves photosynthesise, and some of the food produced is stored in the bulb ready for the winter, while some is used to swell the axillary buds of the bulb. These then produce new bulbs — a method of vegetative reproduction. The green leaves die down later in the year, and the bulb then remains dormant through the winter.

Q. 11:

Give an illustrated description of the development from the egg to the adult of a NAMED insect which shows complete metamorphosis. List the ways in which the development of the frog differs from your example.

[*It is necessary here to define a complete metamorphosis, and then show that your example fits the definition. There is no need to write a description of the features that are shown in the illustrations. Notice that the last part of the question asks for ways in which the developments differ. Do not give a comparison of the stages of development.*]

A: Metamorphosis in insects is a fundamental change in the form of the body during the development from larva to adult. The adult bears little resemblance to the larva. If this change is gradual the metamorphosis is incomplete, *e.g.* in the Cockroach or Frog; but if there is a distinct developing phase, the metamorphosis is complete.

The Large Cabbage White Butterfly is an insect which undergoes complete metamorphosis.

The eggs of the Cabbage White Butterfly are laid in batches of ten to a hundred on the underside of the leaves of cruciferous plants, during the summer. They hatch in about a week, and the larvae emerge.

The larvae of Butterflies and Moths are called caterpillars. The caterpillar of the Cabbage White Butterfly is light green with three longitudinal yellow lines running the length of the body. It is slightly hairy. The head bears a pair of small antennae, six pairs of simple eyes, and a spinneret for spinning silk. The mouth-parts on the lower side of the head are simple mandibles.

The larva is the active feeding and growing stage of the life-cycle, and as it grows it passes through five *ecdyses* or moults, casting its rigid skin each time to allow further growth. After its last ecdysis the caterpillar leaves its food plant and climbs a support, usually a gate or fence, where the pupa is formed.

The pupa is called a *chrysalis* and is attached to its support by a silk pad and girdle formed by the caterpillar's spinneret. The skin of the caterpillar is shed, and the pupa is enclosed in a thin yellow skin through which the developing organs of the adult butterfly can be seen. This is the developing stage. The pupa does not feed, but there is a complete reorganization of the larval tissue, which is broken down and reformed into the tissues of the adult. In the early summer, the development from larva to adult takes about seventeen days, but if the chrysalis is formed during the late summer, it overwinters and the adult emerges in the following April.

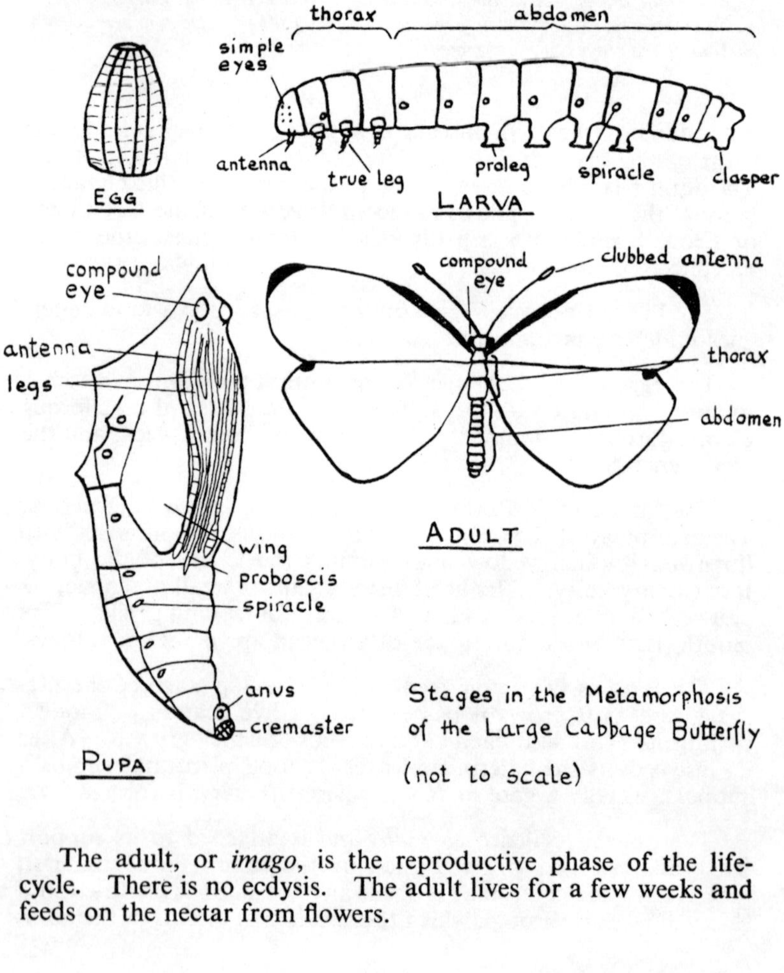

Stages in the Metamorphosis of the Large Cabbage Butterfly

(not to scale)

The adult, or *imago*, is the reproductive phase of the life-cycle. There is no ecdysis. The adult lives for a few weeks and feeds on the nectar from flowers.

The metamorphosis of the Frog differs from that of the Butterfly in the following ways:—

(i) it is gradual, while that of the Butterfly is in distinct stages;

(ii) some of the changes in the Frog, such as the loss of gills and the development of lungs, are related to changes of habitat;

(iii) the Frog takes several years to reach sexual maturity after metamorphosis is complete, but the Butterfly is sexually mature at the end of metamorphosis.

Q. 12:

Name FOUR animals to be found in a garden soil and give an account of their activities. Indicate whether or not they are of use to the gardener.

[*A straightforward question, with a very large number of examples from which to choose. In making the selection, it is advisable to choose at least one harmful and one beneficial type.*]

A: The Earthworm, Centipede, Wireworm and Leatherjacket are all found in garden soil.

EARTHWORM

Earthworms are found in the top layers of most soils which contain sufficient moisture and humus. If the soil becomes dry or excessively cold they migrate to the lower layers where they hibernate during the winter in rounded chambers. On damp nights earthworms leave their burrows in search of food, often keeping the posterior ends of their bodies in their burrows so that they may retreat hastily if threatened. Worms feed on dead vegetable matter which they frequently drag into their burrows; this adds humus to the soil and is therefore beneficial to the gardener.

The burrows are formed by the worms forcing and eating their way through the soil, compressing the walls with their flattened tails, and cementing them with the slime from their bodies. The finely ground soil is passed out of their bodies to the surface as worm casts. Worm casts are a nuisance on lawns, but they bring to the surface some of the deeper soil, covering it with a fine tilth which is excellent for plant growth. The burrows help to loosen, aerate and drain the soil.

Centipede

The centipede found in garden soil is about 3 cm. long and orange-yellow in colour. It is carnivorous, and like other carnivores is very active. It has biting mandibles and a pair of poison claws near the mouth; the claws are used to kill other animals living in the soil.

There is a pair of legs on each of the body segments, and when the animal moves a wave-like motion travels down the line of legs on either side of the body.

The sexes are separate, and the eggs are laid singly in the soil, and hatch out as small centipedes with seven pairs of legs.

Centipedes are of help to the gardener, as they eat pests in the soil.

Wireworm

The wireworm is the larva of the Click Beetle. The eggs are laid during the spring in batches of 50 – 100 and hatch out in July as small golden-yellow wireworms. They take five years to reach their full size of about 2 cm. long. They pupate at the end of July, and the adults emerge in the following spring.

Wireworms can do a great deal of damage to garden crops by eating through the plants just below ground level. They can be controlled by dusting the soil with insecticidal powders, at the risk of disturbing the general ecology.

Leatherjacket

Leatherjackets are the larvae of Crane Flies. They are large, sluggish, legless larvae with a dull-brown leathery integument. The eggs are laid in September in batches of about 200, and the young larvae hatch out in just over a week. They remain quiescent during the winter, and begin feeding in the following spring. Pupation takes place in July or August, and the adults emerge at the end of August.

Leatherjackets are found in damp soil where they can do great damage to garden crops by eating them off at ground level. They can be controlled by baiting them with bran and Paris Green, or in the same way as wireworms.

Q. 13:

Give an illustrated account of the life-cycle of any NAMED insect. Show how the various stages in the life-cycle appear to be adapted to the conditions under which they live.

A: THE HOUSE FLY

The female house fly lays her eggs in batches of somewhat over one hundred in warm, moist, organic material like dung or rotting refuse. The eggs are laid a week after mating and take about one day to hatch. They are about 1 mm. long and white.

The white, legless larva is negatively phototaxic and burrows into the food material, which it digests outside the body by secreting enzymes into it, so that the food is taken in as a liquid. The conical, segmented larva has spiny pads on segments 6 to 12 to assist movement. The mouth-parts are reduced to a single hook which the larva uses when dragging itself through its substrate. There is a pair of spiracles on the posterior end, and a pair on the third segment of the second and third instars.

The larva moults twice in the three to sixty days that it takes to reach its full size (10 mm.), and the pupa forms inside the larval skin, which is then called a puparium. Pupation takes place away from the food source, frequently in soil, and the adult emerges after four to five days.

The young adult has a bladder (*ptilinum*) on the head which is filled with blood and then emptied. This process is repeated until it bursts the puparium from which the imago emerges. The ptilinum then disappears.

The adult rests on the surface of the soil until the wings and exoskeleton become rigid. There is one pair of wings, the second pair of the typical insect being represented by the halteres which act as stabilizers during flight. The halteres are covered by the plate-like squamae. There are three pairs of jointed legs. The head bears a pair of compound eyes, and three simple eyes (ocelli). The antennae are short, and the mouth-parts are modified as a sucking proboscis. The adult feeds on food that has been liquefied by its saliva.

The fly does not achieve sexual maturity until a fortnight after it has emerged from the pupa. Both pupae and adult flies may survive the winter.

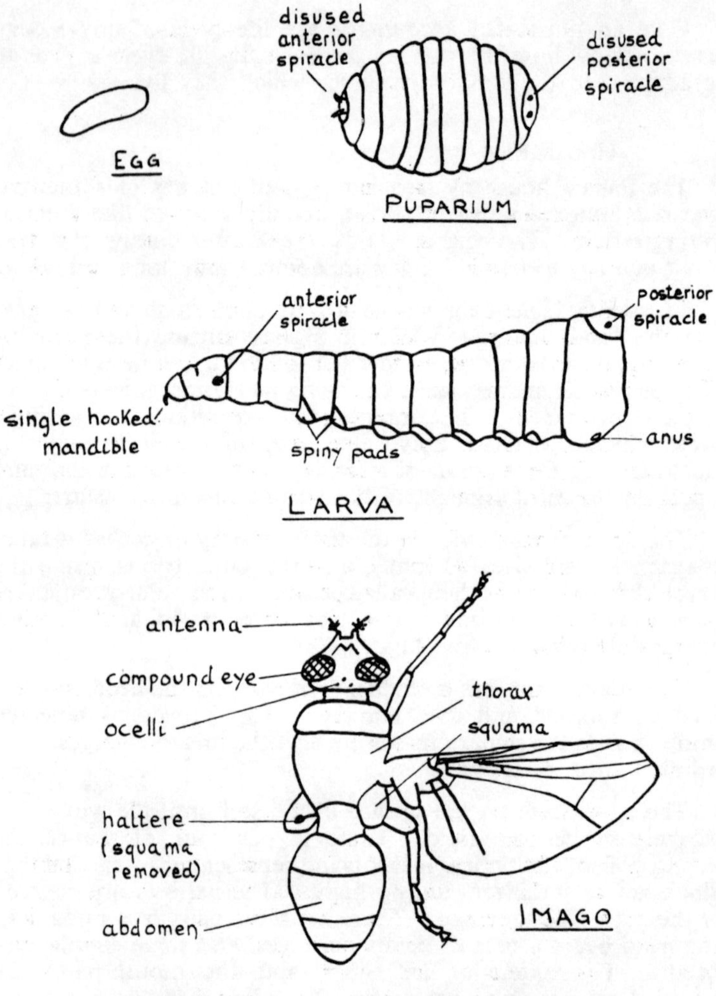

THE HOUSE FLY

ADAPTATION TO ENVIRONMENT: LARVAL STAGE

(i) As the larva lives in intimate contact with its food it needs no legs to move to its food, eyes to see it, or complex mouth-parts to capture and dismember it. These organs are therefore absent or vestigial. The hooked mouth-part and the pads are adequate locomotory organs for slow progress through the food material.

(ii) The larva's negative phototaxis drives it into the darker and moister parts of the substrate and away from the surface, which affords no protection and which may become too dry. When the rest of its body is embedded, the larva can breathe through its posterior spiracles, which remain exposed to air.

PUPA

(i) The last larval skin forms a protective cuticle which prevents desiccation; its dark brown colour renders the pupa inconspicuous.

IMAGO

(i) The mouth-parts are highly specialized in relation to the adult's diet.

(ii) The large compound eyes give good all-round vision which helps the animal to escape quickly from predators.

(iii) The halteres, acting as stabilizers, aid quick, accurate flight.

[*You may use any insect, but it is advisable to quote an example which will give plenty of scope in answering the second part of the question. The Cockroach would not be a good example, because the changes from egg to adult are gradual.*]

Plant Physiology

Q. 14:

What do you understand by respiration? Describe carefully how you would demonstrate that a green plant respires.

[*The definition of respiration must be applicable to both plants and animals. Some aspects of respiration are difficult to demonstrate in a large green plant, e.g. the release of energy, so in answering this question it is advisable to refer to small seedlings such as cress seedlings.*]

A: Respiration is the enzymatic breakdown of complex chemicals to release energy. Respiration takes place in every living cell, usually involving the oxidation of stored or digested food substances by oxygen. In the case of a carbohydrate, carbon dioxide, water and energy are released:—

$$C_6H_{12}O_6 + 6\ O_2 = 6\ CO_2 + 6\ H_2O.$$

Most tissues are capable of at least a limited amount of anaerobic respiration, while some bacteria are wholly anaerobic. Anaerobic respiration does not release energy as efficiently as aerobic respiration:—

$$C_6H_{12}O_6 = 2\ C_2H_5OH + 2\ CO_2.$$

Apart from the chemical breakdown of food, the term respiration is often applied to the mechanism whereby the organism obtains oxygen, *e.g.* by breathing in Man, or diffusion through the stomata and lenticels in the flowering plant.

To show that aerobic respiration takes place it is necessary to prove that oxygen is used, carbon dioxide is given off, and that there is a loss of stored food, *i.e.* a loss of weight.

Green plants suitable for such an experiment are germinated cress seedlings.

A known number of seedlings are placed in the apparatus shown. The retort is blackened to avoid the removal of carbon dioxide by photosynthesis. After a few days, the caustic soda solution will have risen up the stem of the retort to occupy one-fifth of the total volume. As one-fifth of the atmosphere is oxygen, and carbon dioxide is the only gas which could have been absorbed by the caustic soda solution, this suggests that oxygen has been used up, and replaced by carbon dioxide. Confirmation may be obtained by showing that the gases left in the retort will not support the flame of a burning taper.

A similar, empty retort is set up as a control in which there is no significant change.

The seedlings are removed from the retort, and oven-dried at 100° C. until there is no further change in weight. The same number of ungerminated cress seeds (about 40) are treated in the same way. On comparing the weights of the two groups, it will be seen that the germinated seeds weigh less than the ungerminated, showing that there is a loss of weight during respiration. Dry weights are used because the germinating seeds will have absorbed water during the initial stages of germination.

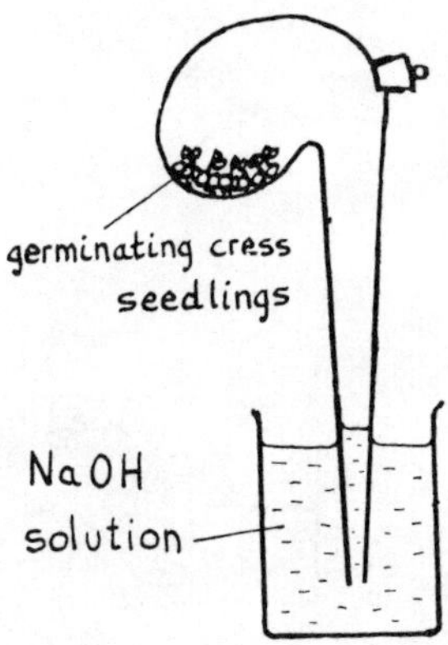

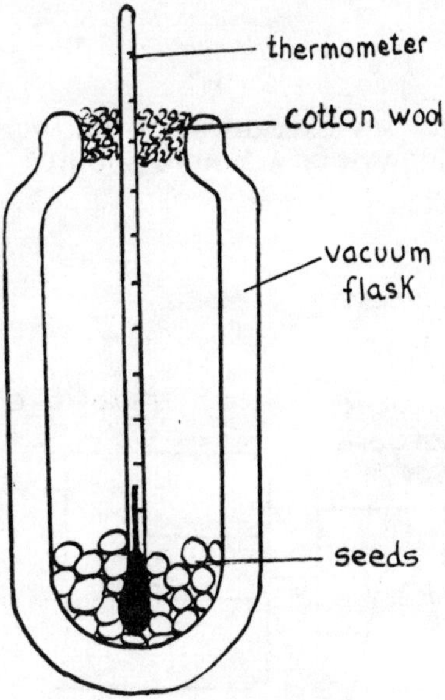

It is possible to show the evolution of energy as heat, in the following way:—

Germinated cress seedlings are placed in the apparatus shown above. A similar flask is set up containing seedlings killed by heating, and surface-sterilized with 0·1% mercuric chloride. After two or three days, the temperature in the first flask rises a few degrees above that in the second, showing that heat is evolved during respiration.

Q. 15:

Describe an experiment to show the effect of gravity on the growth of a young shoot. Explain as far as you can the processes involved in this phenomenon.

A: AN EXPERIMENT TO SHOW THE EFFECT OF GRAVITY ON THE GROWTH OF A YOUNG SHOOT.

APPARATUS:

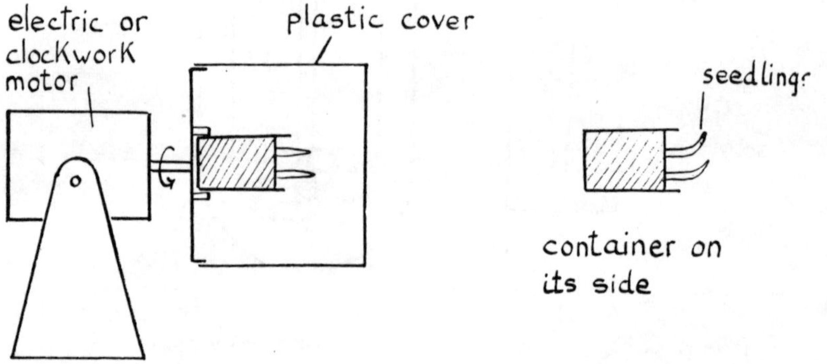

METHOD: Four groups of Oat seedlings are grown in soil in small tubes which will fit into the spring-holder of a clinostat. When the coleoptiles have emerged, the tips are cut off the seedlings in two of the groups. One group of 'cut' seedlings and one group of 'whole' seedlings are placed on clinostats as shown in the diagram, and revolved at four revs./min., so that gravity can affect all sides of the plants equally. The other two groups of plants are put on their sides. The whole experiment is carried out in the dark to eliminate the effect of light.

OBSERVATION: After a few days, all the seedlings will have grown, but only the 'whole' seedlings on their sides will have bent upwards, *i.e.* will have shown a negative geotropism. The other three groups will continue to grow straight.

CONCLUSION: When 'whole' plants are influenced by gravity equally on all sides, they grow straight; but if the influence of gravity is unilateral, they grow away from it. This effect is brought about by the shoot tip affecting the growing region behind it. This is shown by the fact that shoots from which the tips have been removed are not influenced by gravity.

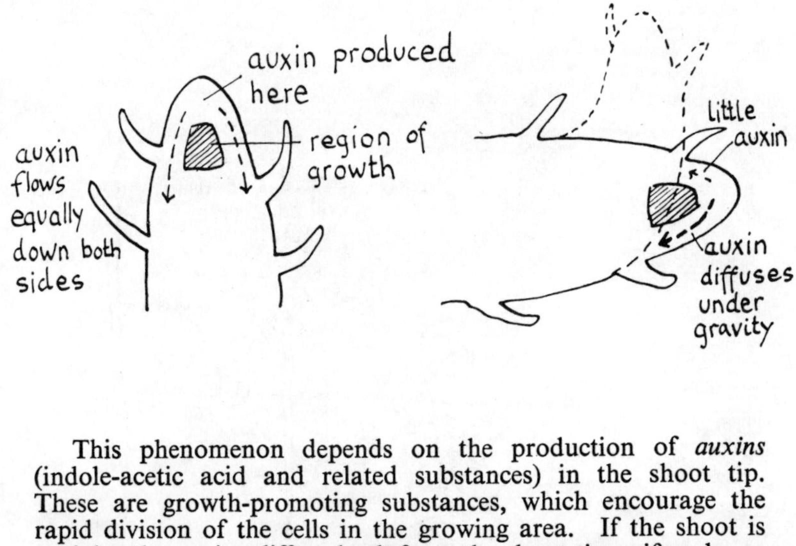

This phenomenon depends on the production of *auxins* (indole-acetic acid and related substances) in the shoot tip. These are growth-promoting substances, which encourage the rapid division of the cells in the growing area. If the shoot is upright, the auxins diffuse back from the shoot tip uniformly, so that all the cells in the growing zone will receive an equal stimulus and divide at the same rate, resulting in straight growth of the shoot. If, however, the plant is on its side, the auxins will tend to gravitate towards the lower surface so that the lower side will receive more auxin and tend to grow at a faster rate than the upper side. The tip of the shoot will consequently curve upwards. Once the tip is upright, the auxin will diffuse equally down all sides, and growth will be vertical.

Q. 16:

Explain with the aid of labelled diagrams the difference between hypogeal and epigeal germination. How would you investigate experimentally the effect of (*a*) ether vapour and (*b*) lack of oxygen on the germination of seeds?

[*Note that the first part of the question does not demand all the stages in germination, so only two diagrams of each type are necessary. Remember that the hypocotyl is the part of the plant between the radicle and the cotyledons, the plumule emerging above the cotyledons.*

The second part of the question asks for the effect of two conditions on germination. It is therefore legitimate to have one control and the two variables in one experiment.]

A:

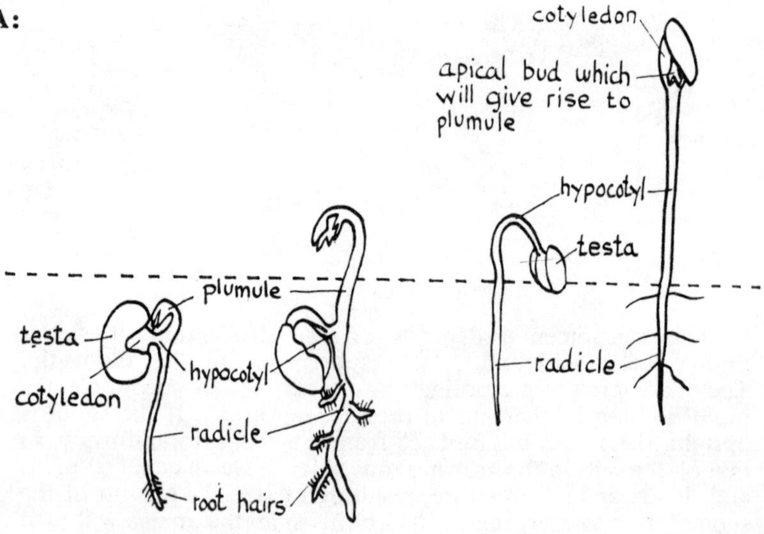

BROAD BEAN: HYPOGEAL GERMINATION

CRESS: EPIGEAL GERMINATION

Hypogeal germination takes place in seeds which have food reserves stored in the cotyledons, and the cotyledons remain underground when the plumule emerges. The first leaves to be produced above the ground are the true leaves. The hypocotyl remains short, and the plumule grows out from the seed.

Seeds showing epigeal germination have their food reserves stored in the endosperm. The cotyledons are carried above the ground as the hypocotyl elongates. The plumule is produced later above the cotyledons, which become green on exposure to sunlight, and are the first photosynthetic organs.

An Experiment to Show the Effects of Ether Vapour, and Lack of Oxygen, on the Germination of Seeds.

Apparatus:

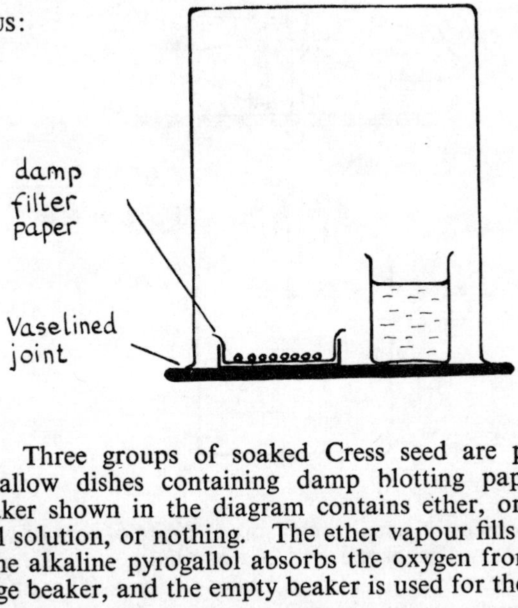

Method: Three groups of soaked Cress seed are placed in small, shallow dishes containing damp blotting paper. The small beaker shown in the diagram contains ether, or alkaline pyrogallol solution, or nothing. The ether vapour fills the large beaker, the alkaline pyrogallol absorbs the oxygen from the air in the large beaker, and the empty beaker is used for the control.

Observation: After a few days the seeds in the control apparatus germinate, but those in the rest of the experiment do not germinate.

Conclusion: The experiment indicates that oxygen is necessary for germination, and that seeds do not germinate in ether vapour.

Q. 17:

Make a large labelled diagram of a potometer. For what purpose is a potometer used? Explain concisely the role of the stomata and root-hairs.

[*This is a question on transpiration in its various aspects. Point out the disadvantages of the potometer. The discussion of the role of the stomata and root-hairs must include some mention of their structure, and the way in which they function.*]

A:

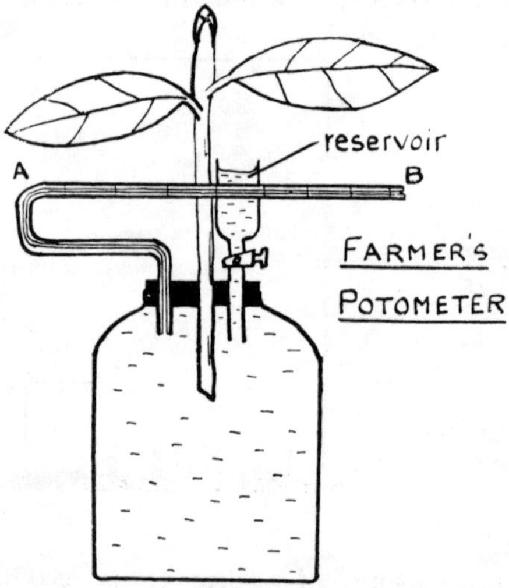

A potometer is used to measure the rate of loss of water from a cut shoot. The bottle and the capillary tube AB are full of water. The reservoir also contains water. The apparatus is placed under various climatic conditions, *e.g.* in a draught, or in still air. Water is lost from the leaves on the shoot, and consequently from the bottle. This loss is shown by the column of water retreating along the capillary. The rate of water-loss can be measured by timing this movement. The capillary tube can be refilled by releasing water from the reservoir into the bottle.

This apparatus is useful only for comparative measurements. The shoot should be cut under water, to avoid the formation of air locks in the xylem. In spite of all precautions the initial readings will be irregular due to a shock reaction of the shoot. The apparatus should be left for a few hours for the shoot to recover before the readings are taken as a measure of transpiration.

The stomata are found on the lower surface of dicotyledonous leaves, and on both surfaces of monocotyledonous leaves. The stomata are pores bounded by a pair of reniform guard cells, which have thickened walls and contain chloroplasts. The function of the stomata is to control the rate of gaseous exchange of the leaves. The stomata open and close, controlling the

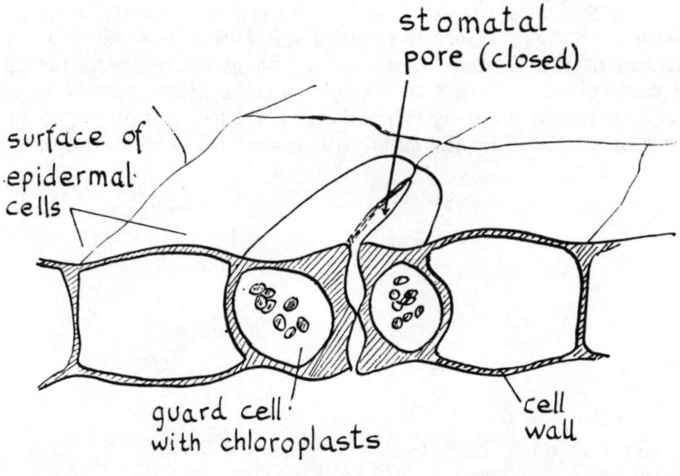

diffusion of carbon dioxide, oxygen and water vapour. During the day the chloroplasts of the guard cells produce sugar by photosynthesis, building up the concentration of the cell sap above that of the cells of the surrounding epidermis, which do not have chloroplasts. Water then passes into the guard cells by osmosis, so that they become more turgid and rounded; this alteration of shape causes the guard cells to separate and leave an open pore. This allows gaseous exchange to take place. At night there is no photosynthesis. The guard cells lose their sugar by respiration so that the concentrations of their cell saps are reduced; water passes out of the guard cells, they become less turgid, and the pore is closed. Under hot, dry, or windy conditions, water is lost more rapidly by transpiration than it is absorbed by the roots. The cells of the leaves, including the guard cells, become flaccid. The guard cells close in consequence, reducing further water-loss from the leaves.

The root hairs absorb water and minerals from the soil. They are projections of the piliferous layer extending into the soil between the particles. The soil solution is more dilute than the cell sap. The cellulose cell walls are completely permeable, but the protoplasmic linings of the cells are semi-permeable. Under these conditions osmosis takes place, and water passes from the soil into the root hairs. Salts pass into the root hairs by ion-exchange. The salts in the soil and in the cell sap are in solution in the form of electrically charged ions. In this case, the protoplasmic linings of the cells act as filters, letting in ions of one charge needed by the cells, and releasing ions of the same charge not needed by the cells, thus maintaining the ionic balance of the plant.

Q. 18:

Describe experiments which would enable you to demonstrate TWO of the following: (*a*) the importance of nitrates in plant growth, (*b*) the liberation of heat in the germination of seeds, (*c*) the path followed in the plant by the water absorbed by the roots, (*d*) the effect of light on the direction of growth of shoots.

A: (*a*) An Experiment to Show the Importance of Nitrates in Plant Growth.

Apparatus:

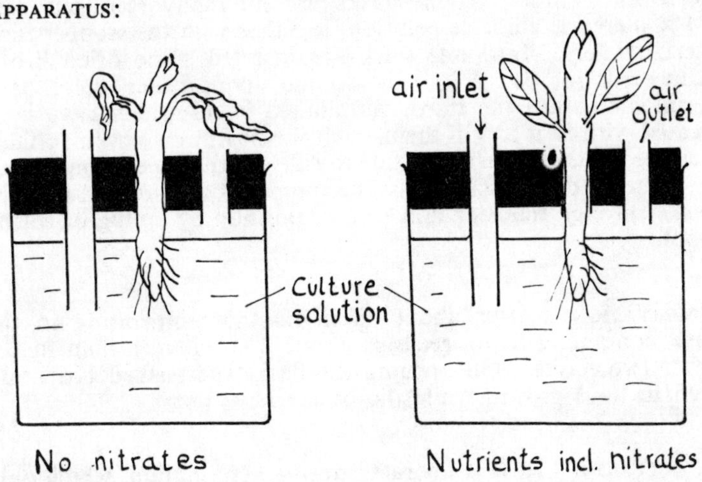

Method: The jars, stoppers and tubes are sterilized. A cutting of an easily rooted plant such as *Tradescantia* is inserted through the central hole in each jar. The jars are covered with dark paper to keep light from the roots and to prevent the growth of algae. One jar is used as a control, and is filled with a culture solution containing all essential nutrients; the second jar is filled with a solution similar in all respects except that the nitrates have been replaced by sulphates. The culture solutions are aerated daily.

Observation: The cuttings root in both jars, but after several days the plant without nitrate ceases to grow normally, and eventually it dies. The plant in the control grows well.

Conclusion: Nitrates are necessary for healthy growth of the plant.

(*b*) AN EXPERIMENT TO SHOW THAT HEAT IS LIBERATED BY GERMINATING SEEDS.

METHOD: Three vacuum flasks are internally sterilized with 0·1% mercuric chloride solution, together with the stoppers and thermometers. Into one flask are inserted some dried broad bean seeds (or peas); into the second, living beans which have germinated; into the third, germinated beans which have been treated with heat to kill them. All the beans have been surface-sterilized with mercuric chloride to rule out the generation of heat by bacterial decomposition. Thermometers are inserted into the flasks through the neck and held in position by a plug of cotton wool.

OBSERVATION: After about 48 hours the temperature in the flask containing living seeds is about 2° C. higher than in the other two flasks. On opening the flasks the soaked beans are seen to have grown, while the others have not.

CONCLUSION: Heat is liberated during germination, which only takes place if the seed has been soaked, and no heat is formed by any other reactions than those associated with germination.

[*The experiments for* (*a*), (*b*), *and* (*d*) *must have controls because it is the presence or absence of a factor which is being investigated.* (*c*) *does not need a control because the result is self-evident if a physiologically inert die is used.*]

Q. 19:

State the composition of the atmosphere. Describe the uses made by plants and animals of the gases present.

A: The composition by volume of the dry air in the atmosphere at ground level is virtually constant at 21% oxygen, 78% nitrogen, 0·03% carbon dioxide, and traces of argon, hydrogen, neon, helium, krypton and xenon. The atmosphere also contains a variable amount of water vapour. Atmospheric hydrogen is not used by living organisms, nor are the other trace gases, which are chemically inert.

Oxygen is used by both plants and animals during aerobic respiration. Respiration takes place in every living cell, where complex chemicals are broken down enzymatically into simpler substances with the release of energy, as in the aerobic respiration of glucose:—

$$C_6H_{12}O_6 + 6\ O_2 = 6\ CO_2 + 6\ H_2O + \text{energy}$$

This type of respiration is called *internal* or *cellular*. The removal of carbon dioxide and water, and the renewal of the oxygen by the organism is called *external* respiration or *breathing*. The method of exchange differs in plants and animals.

In plants, external respiration takes place by diffusion, through the general surface in simple plants, *e.g.* Spirogyra, and through inter-cellular cavities connected to the outside by simple pores (stomata and lenticels) in the more complex plants.

Gaseous diffusion through the body surface can also supply sufficiently small animals with oxygen. This method has the disadvantage that the body surface has to be kept wet, as the gases have to be dissolved. When the volume/surface-area ratio is increased, some method of increasing the respiratory surface without further increasing the body volume is necessary. Various types of organ achieve this, *e.g.* the trachea of insects, the gills of fish, and the lungs of man. In all terrestrial animals, the surface for gaseous exchange is inside the animal so that it can be kept moist. Frequently diffusion does not transport enough air to the respiratory organs and it becomes necessary to have some form of pumping mechanism, *e.g.*, in man, the inhaling and exhaling of air by using the diaphragm and intercostal muscles with the ribs.

Carbon dioxide is used by green plants during photosynthesis. In the presence of chlorophyll, light, and an adequate water supply, a living cell can build up water and carbon dioxide into a monosaccharide sugar, releasing oxygen. Some of this oxygen is used in respiration. This is an anabolic process storing the light energy from the sun in the form of chemical energy in the sugar molecule:

$$6\,CO_2 + 6\,H_2O = C_6H_{12}O_6 + 6\,O_2$$

Plants also use carbon dioxide in the formation of insoluble carbonates which are excretory products. These are stored as crystals, and are usually removed during leaf fall.

There is no anabolic process in animals which uses atmospheric carbon dioxide directly, but in its dissolved form of carbonic acid the gas plays an important part in various processes in animals. A large proportion of the carbon dioxide is produced during respiration. Such processes include the manufacture of sodium carbonate and bicarbonate, used to make the digestive juices of the mammal alkaline, *e.g.* the bile. The production of the excretory product urea begins with the formation of ammonium carbonate, which involves carbon dioxide. Calcium carbonate is found in the shells of molluscs *etc.*

Atmospheric water vapour is not usually taken in by plants and animals, but it controls the rate of loss of water by an organism. A humid atmosphere reduces the rate of transpiration of plants; it also reduces the evaporation of sweat of a mammal, thus lowering the efficiency of the body cooling system.

The only organisms to make direct use of the nitrogen in the air are the nitrogen-fixing bacteria, which can combine it with carbohydrates to make amino-acids and hence proteins. Through them, atmospheric nitrogen becomes available to plants and animals.

Animal Physiology

Q. 20:

Explain how (*a*) oxygen intake, (*b*) excretion and (*c*) food capture are carried out in an Amoeba. In what ways does an Amoeba differ from a single cell of Spirogyra?

A: (*a*) Oxygen intake in Amoeba is by direct diffusion through the ectoplasm, and through the surfaces enclosing the contractile vacuole and other water-filled vacuoles, of oxygen dissolved in the water in which the animal lives.

(*b*) Amoeba excretes the waste products of its metabolism by diffusion through the cell membrane, either directly to the outside, or from the endoplasm into vacuoles which discharge into the surrounding water. Some waste matter never enters the cytoplasm at all, but remains in the food vacuole in which it has been formed until the vacuole is emptied. The contractile vacuole probably aids excretion (and respiration) only indirectly by regulating the water content of the cell body.

(*c*) Amoeba feeds on small organisms such as algae, bacteria and organic debris by engulfing them with pseudopodia. The diagram shows how this leads to the formation of a food vacuole. Although the surfaces of the pseudopodia consist of ectoplasm, the food vacuole is either passed into the endoplasmic region or (more probably) the cytoplasm enclosing it reverts to the endoplasmic state. Amoeba is aided in the capture of food by its chemical sensitivity, and by a certain 'sticky' property of its pseudopodia which makes it difficult for small ciliates to escape once they are in close contact with a pseudopodium.

Amoeba differs from a single cell of Spirogyra in a number of fundamental ways. The body of Amoeba is of indeterminate shape, while that of Spirogyra is enclosed in a cylindrical cellulose wall. Amoeba is motile, holozoic and contains no chlorophyll; Spirogyra has no powers of independent motion, feeds holophytically, and contains chlorophyll in its chloroplasts so that it can use the energy of sunlight for photosynthesis. These differences place Spirogyra firmly in the Plant Kingdom, and Amoeba in the Animal Kingdom.

In Amoeba the cytoplasm contains a number of vacuoles; of Spirogyra it is truer to say that the vacuole contains most of

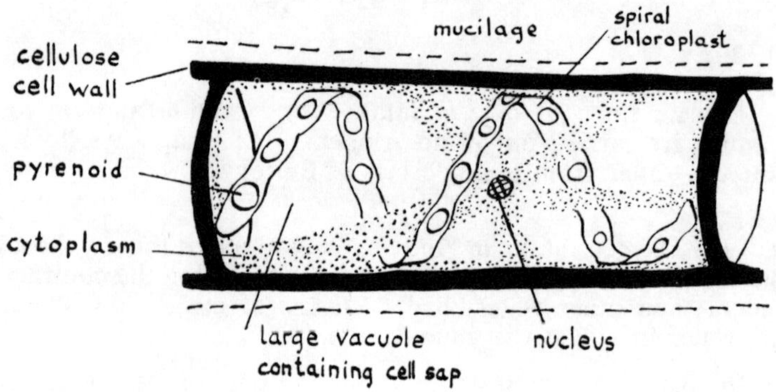

SPIROGYRA

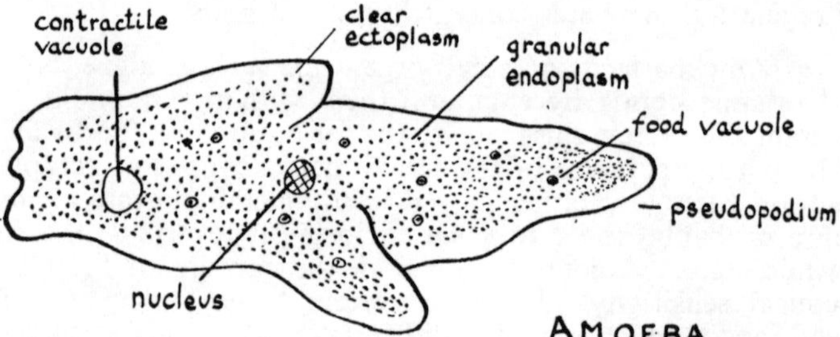

AMOEBA

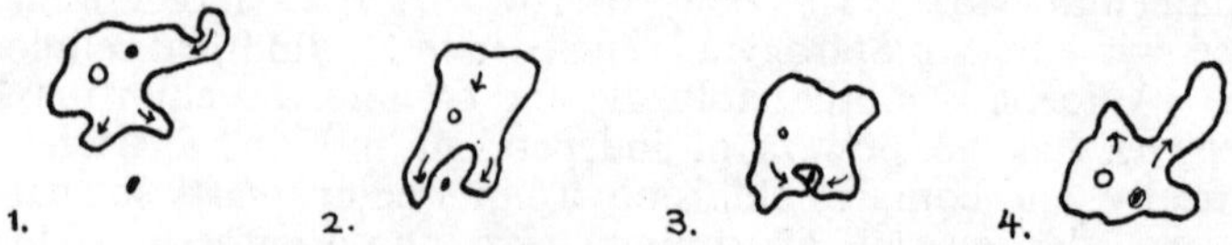

Stages in ingestion of food particle by Amoeba

the cytoplasm. There is no contractile vacuole in a Spirogyra cell. Amoeba is a complete, non-cellular organism; Spirogyra is multi-cellular, and although all its cells are similar and capable of independent existence, a single cell is not representative of the species. Amoeba can encyst; Spirogyra cannot.

[*Note that the question asks only for differences between Amoeba and a single cell of Spirogyra; there is therefore no need to discuss reproduction, which in Spirogyra involves more than one cell.*]

Q. 21:

Explain, with the aid of labelled diagrams, how the structure of the thorax of a mammal enables air to be drawn into the lungs. State the differences between the inspired and expired air, and account for these differences as far as you can.

A: In the mammalian thorax, the lungs are contained within two airtight compartments known as the *pleural cavities.* There is little or no air in these cavities, and each of the two lungs fills its cavity almost completely. When the volume of the thorax is increased by muscular action, the outer pleura tends to pull away from the inner pleura and so to create a vacuum in the pleural cavity. The lungs are open to the air through the trachea and the mouth or nose, and as long as the atmospheric pressure is greater than that in the pleural cavity, air streams into the lungs, thereby expanding them and keeping the cavities filled. When the thorax relaxes, the volume of air that can be accommodated in the lungs is reduced, and the excess is expelled via the trachea. By the repetition of this cycle of inspiration and expiration a supply of fresh air to the lungs is maintained.

The volume of the thorax is increased in two ways: (*a*) by contraction of the muscles of the diaphragm, which in the relaxed position bulges upwards into the thorax; and (*b*) by expansion of the rib cage through contraction of the inspiratory intercostal muscles. During expiration, these muscles relax, and in addition the abdominal wall contracts, driving the stomach and liver against the diaphragm and making it bulge into the thorax again.

On comparing the percentage composition of exhaled air with inhaled air, it is found that oxygen has dropped from 20%

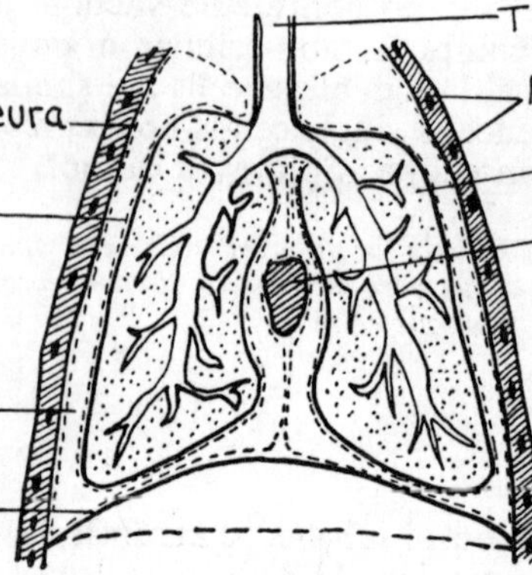

to about 15%, carbon dioxide has increased from 0·03% to almost 4%, and water vapour from about 1% to 6%. The increase in water vapour is accounted for by the passage of the air past the mucuous membranes and moist surfaces of the entire respiratory tract. The other two changes are due to an exchange of gases with the blood which takes place by diffusion in the alveoli of the lungs.

The blood which is conveyed to the lungs direct from the heart through the pulmonary artery is in fact venous blood with a concentration of oxygen in it which is lower than that in the air-sacs of the lungs. Oxygen therefore diffuses through the walls of the alveoli into the capillaries and enters into loose combination with the pigment haemoglobin of the red blood corpuscles. This accounts for the 5% drop in oxygen.

The re-oxygenated blood is returned to the heart and then pumped to the body tissues generally. The energy requirements of active tissues are met by combustion of organic substances, a process which takes up oxygen and releases carbon dioxide. The blood surrenders oxygen wherever the concentration of the latter is low, and carbon dioxide is taken into the plasma by solution or formation of salts wherever its concentration is high. When the blood reaches the lungs after its next passage through the heart, the concentration of carbon dioxide in it is higher than in the alveoli, and some of the gas diffuses out of the blood into the air-sac. This accounts for the higher carbon dioxide content of expired air.

Q. 22:

Explain concisely the functions of a nervous system. With the aid of a large, labelled diagram, give a concise account of the ear of a mammal.

A: The basic function of a nervous system is to transmit information or 'instructions' in the form of electrical impulses from one part of an organism to another. In its most rudimentary condition, as in Hydra, the nervous system consists simply of a network of branching nerve cells. In higher animals, four types of structure are distinguished corresponding to four functions:—

(*a*) the sense organs or receptors, whose function is to detect light, heat, cold, sound, pressures, and chemical substances, in the external or internal environment.

(*b*) the afferent nerves, whose function is to convey signals arising from the stimulation of the sense organs to the central nervous system.

(*c*) the central nervous system, comprising the brain and spinal cord in vertebrates, in which incoming information is collated and, to a varying extent, memorised, and in which outgoing impulses are determined.

(*d*) the efferent nerves, which transmit nervous impulses from the central nervous system to the muscles, organs or glands from which a response is required (the effectors).

An essential property of nervous systems, and one which contrasts with the slow mechanisms of chemical conductivity found in plants, is that they act almost instantly and can produce quick responses to a large variety of circumstances.

The mammalian ear is a sense organ concerned with hearing, and with posture and balance.

In the outer ear, the *pinna* and the external auditory passage collect and conduct sound waves to the ear drum (tympanum). The latter transmits its vibrations through the three small, linked bones in the middle ear (malleus, incus and stapes) to the membrane of the oval window (fenestra ovalis) in the inner ear. The

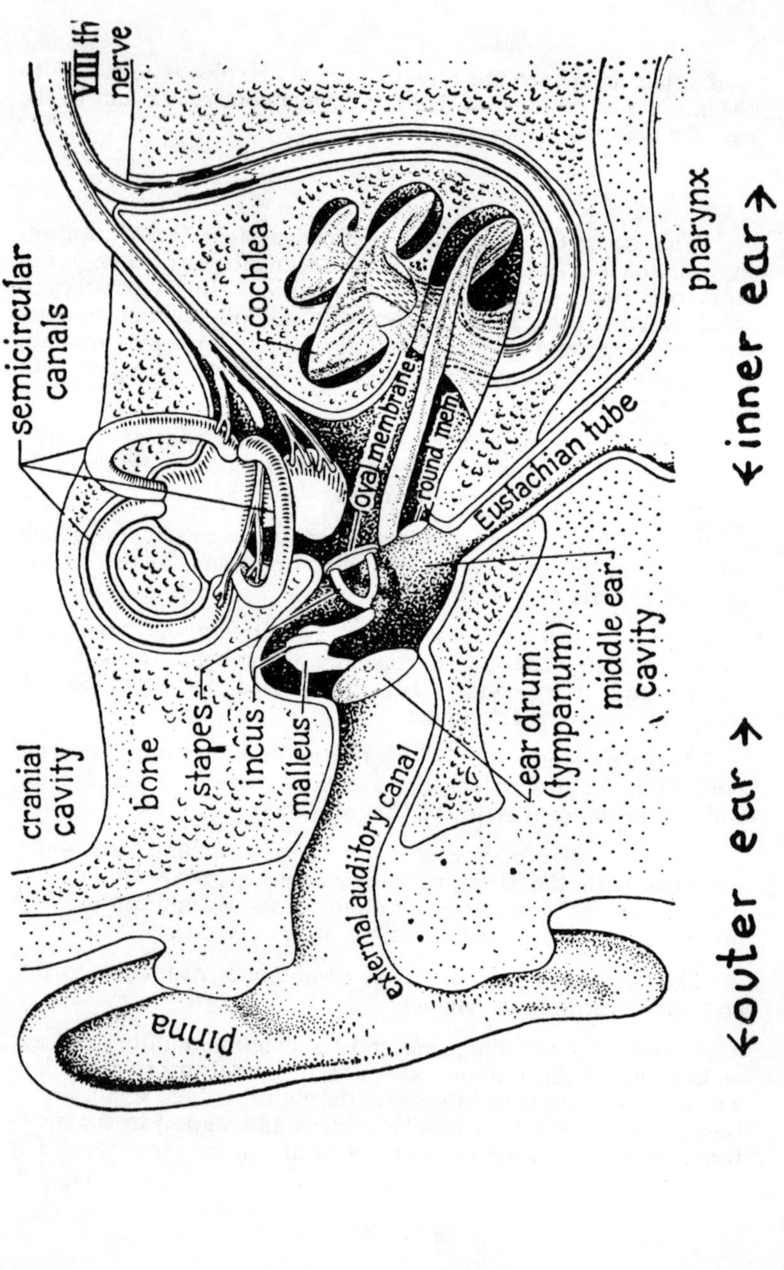
VIII th nerve
semicircular canals
cochlea
oval membranes
round mem.
Eustachian tube
pharynx
cranial cavity
bone
stapes
incus
malleus
external auditory canal
ear drum (tympanum)
middle ear cavity
pinna
← outer ear →
← inner ear →

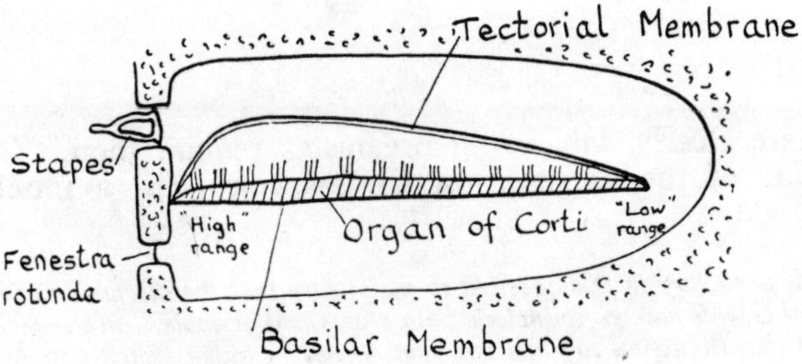

Diagrammatic Section of uncoiled Cochlea.

Eustachian tube connects the middle ear with the mouth cavity and intermittently equalises the atmospheric pressure on both sides of the tympanum.

The inner ear or labyrinth is filled with fluid. Vibrations reaching the oval window travel through the fluid to the cochlea and make the basilar membrane in it resonate. This resonance sets into motion certain hairs of the organ of Corti according to the frequencies present in the vibration. The further ends of the hairs are embedded in the tectorial membrane, in which nerve cells detect the movements of the individual hairs and transmit them in detail through the auditory nerve to the brain.

The semi-circular canals in the labyrinth are at right angles to each other, and any rotation of the head sets the fluid in at least one of them into motion relative to the walls of the canal. This flow bends the hairs of the *ampulla*, which project into the fluid, and nerve endings in the hair cells transmit information to the brain. Similar devices in the trunk of the labyrinth convey details of acceleration and inclination of the head to the central nervous system.

[*Let the diagram show the structure of the ear, and the text explain its working. Do not duplicate details by giving them in both the text and the diagram.*]

Nutrition

Q. 23:

State clearly why living organisms require food. Give an account of the nutrition of (i) tapeworm, (ii) a mould and (iii) dodder.

[*Care is necessary in this question to make sure that the discussion of the need for food is wide enough to include both plants and animals. In the second part confine the discussion to nutrition, and do not be side-tracked into remarking on life-cycles.*]

A: Food is required by living organisms for the production of energy, for growth, replacement of used tissue, reproduction and the maintenance of good health. Energy is released by the enzymatic breakdown and oxidation of complex molecules, and the necessary oxygen is supplied by cellular respiration. This energy is needed for building up new complex molecules from simpler ones, for movement, for the energy requirements of the nervous system and, in warm-blooded animals, for the maintenance of the body temperature which permits the metabolic processes to take place at the right rate. Plants need much less food than animals of comparable weight, partly because they can utilise the energy of sunlight and synthesise organic molecules from carbon dioxide and water during photo-synthesis.

Energy is derived mainly from fats and carbohydrates Growth, replacement and reproduction involve the manufacture of protoplasm from proteins and water. This is brought about by the recombination of simpler molecules, derived from the food, into the complex molecules of protplasm. Various salts, *e.g.* those of calcium and iron, are also necessary for these vital processes, and are also derived from the food. Vitamins are necessary in small quantities for certain vital functions, and, while some can be manufactured by healthy organisms, others must be taken in food.

THE TAPEWORM

The adult Tapeworm is an internal parasite in the intestine of Man and other animals. It is bathed in the digested food in the intestine of the host, and absorbs its food over the whole body surface. Because it is soaked in digested food, it does not need a digestive system of its own. The absorbed nutrients are utilized by the tapeworm tissues, mainly in the production of eggs to ensure the survival of the species. Similarly, the immature (bladderworm) stage in the muscles of the Pig absorbs nutrients direct from the body fluids in the muscular tissue.

Mucor: A Mould

Mucor is a saprophytic fungus growing on a wide range of organic substrates, *e.g.* damp bread. The spore germinates, and the hyphae penetrate into the substrate. There is no chlorophyll, and food is obtained by extra-cellular digestion of the substrate. The tips of the hyphae secrete enzymes which break down proteins to amino-acids, and carbohydrates to monosaccharide sugars. Both these foods are absorbed by the hyphae, and are re-synthesised inside them into either new protoplasm or oil droplets for storage.

Dodder

Dodders are parasitic flowering plants invading Nettle, Gorse or Heather, according to the species of Dodder. The seed germinates in the soil, and the young seedling obtains nutrients from the seed. The mature plant becomes attached to its host by *haustoria*, and it is completely parasitic, having no functional green leaves or roots. The haustoria of the parasite develop xylem and phloem which penetrate into the corresponding tissues of the host stem. The dodder absorbs water and salts from the xylem, and elaborated food materials from the phloem. These substances are synthesised in the parasite to make new tissues.

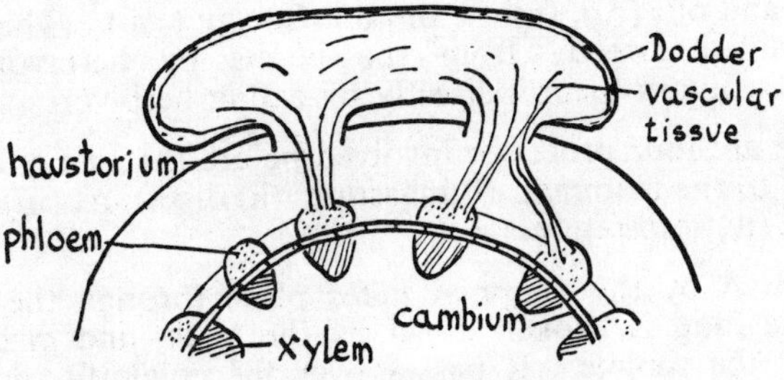

Half T.S. of host stem with Dodder

Q. 24:

Explain how energy from the sun becomes available to a mammal through the agency of the green plant.

A: Green plants convert some of the energy of sunlight into chemical energy during photosynthesis. This energy is used to combine the simple molecules of carbon dioxide and water into complex sugar molecules, and is stored in these larger molecules as chemical energy which is ultimately released during respiration, either by the plant itself, or, in the case under discussion, by the mammal.

The products of photosynthesis are monosaccharide sugars, which may be stored as starch or as disaccharide sugars. The carbon, hydrogen and oxygen combined during photosynthesis are also used in the synthesis of protein. The enzymatic combination with nitrogen and other elements brought up from the soil in the water as salts leads to the production of amino acids, and ultimately protein.

Photosynthesis takes place in a living cell only in the presence of light, chlorophyll, and an adequate supply of carbon dioxide and water. When a monosaccharide sugar, for instance glucose, is formed, the sequence of reactions may be summed up as:—

$$6\,CO_2 + 6\,H_2O = C_6H_{12}O_6 + 6\,O_2$$

The high-energy foods which the mammal may obtain from a green plant include monosaccharide sugars, sucrose (*e.g.* from Sugar Beet), starch (*e.g.* from Potatoes), protein from any protoplasm (leguminous seeds are a particularly good source), and fats and oils (*e.g.* from Palm oil, Coconut oil). These foods are obtained directly from the plants by herbivores, but carnivores obtain them indirectly by eating herbivores.

There are four processes involved before the energy becomes available to the mammal: (i) ingestion, (ii) digestion, (iii) absorption, and (iv) respiration.

Ingestion by the mammal takes place through the mouth, where the food is broken down by the teeth, and mixed with saliva by the tongue. It passes over the epiglottis, down the oesophagus by peristalsis, and through the cardiac sphincter into the stomach. After some time the food passes out through the pyloric sphincter into the duodenum where digestion continues. and ultimately into the ileum where the digested substances are absorbed.

Digestion of starches begins in the mouth of a human with the production of ptyalin, which breaks down starch to maltose. When the food reaches the stomach, the hydrochloric acid increases the acidity to such a level that the ptyalin ceases to function; but under these conditions, the pepsin produced by the stomach breaks down protein to peptones. In the duodenum the food becomes mixed with bile, which is produced by the liver and contains bile salts and sodium bicarbonate. The former emulsifies fats, and the latter makes the gut content (*chyme*) alkaline. The pancreatic juice entering the first loop of the duodenum through the pancreatic duct contains sodium bicarbonate and carbonate, and trypsinogen, which forms trypsin when combined with enterokinase in the intestinal juice. The trypsin breaks down proteins to amino acids. Amylopsin, digesting starches to maltose, and lipase, breaking down fats to fatty acids and glycerol, are also contained in the pancreatic juice. Apart from enterokinase, the intestinal juice contains erypsin, which breaks peptones down to amino acids, sucrase, which turns sucrose into monosaccharide sugars, and maltase, which alters maltose to glucose.

The end products of digestion are therefore monosaccharide sugars, amino acids, and glycerol and fatty acids. These are absorbed through the villi of the ileum, the monosaccharide sugars and amino acids into the blood vessels, and the glycerol and fatty acids into the lymphatic system, which joins the blood system through the subclavian veins.

The digested food is carried by the blood system to the site where it is required; some of the carbohydrates may be stored, as glycogen, in the liver.

Respiration of the digested food then takes place in the tissues, releasing the energy which has ultimately come from the sun. Aerobic respiration with carbohydrates yields carbon dioxide, water and energy:—

$$C_6H_{12}O_6 + 6\,O_2 = 6\,CO_2 + 6\,H_2O$$

This is the direct reversal of the photosynthetic anabolism.

The oxidation of protein yields approximately the same amount of energy as carbohydrate, weight for weight, but ammonia is one of the breakdown products. The respiration of fats produces the same end-products as carbohydrates, but gives more energy, weight for weight.

Although the processes of mammalian digestion and respiration give a net yield of energy, they require a considerable amount of energy themselves, so that the animal benefits from only a portion of the energy of sunlight that was originally stored in green plants.

[*Note: the question asks how the energy becomes available to a mammal, and the answer must therefore give some account of ingestion, digestion, absorption, and respiration, but not of excretion.*]

Q. 25:

Why is digestion necessary? How do (*a*) the salivary glands, (*b*) the stomach, (*c*) the pancreas help in the digestion of food in a mammal?

[*There is a danger in a question like this that the candidate will be side-tracked into a discussion of the whole digestive system.*]

A: Digestion is necessary to extract nutrient substances from crude food in a form in which they can diffuse into the organism. In mammals, the food has to be broken down mechanically and chemically until the nutrients are in solution and the larger molecules have been reduced to smaller ones which can enter the blood-stream by diffusion through the walls of the intestine and the capillaries. Much of the chemical work is done by enzymes.

(*a*) In Man, there are three pairs of salivary glands situated in the mouth. They secrete a solution of mucin and ptyalin. The water begins to dissolve any soluble food, and together with the mucin lubricates the food, helping its passage down the oesophagus to the stomach. The ptyalin digests starch to soluble maltose. In many other animals, the saliva has little or no digestive function.

(*b*) The stomach is a muscular sac, and the rhythmic contraction of the stomach wall mixes the food inside it. The numerous glands of the inner wall secrete hydrochloric acid, pepsin and, in young animals, rennin. The hydrochloric acid acidifies the stomach contents, stopping the action of the ptyalin, but enabling the pepsin and rennin to function. The pepsin breaks down proteins to peptones and proteoses, and the rennin coagulates milk. Having been in the stomach for some time, the food is at body temperature, and the fats are melted.

Many harmful substances, and excessive quantities of normally acceptable food, act as irritants in the stomach, and can in many animals evoke a convulsive upward constriction whereby the stomach contents are vomited.

(*c*) The pancreas is a large, diffuse gland lying in the first loop of the duodenum. Its duct has a common opening with the bile duct into the duodenum. The contents of the duodenum are made alkaline by the bile, and by the sodium carbonate and bicarbonate in the pancreatic juice. Under these conditions, the action of the pepsin is stopped, and the enzymes secreted by the pancreas and duodenum can function.

The pancreatic juice contains lipase which digests fats to fatty acids and glycerol; amylopsin which further digests starch to maltose; and trypsinogen. The action of lipase is aided by the bile, which emulsifies the fats and so exposes a larger area to the action of the enzyme. Trypsinogen is a precursor of trypsin which breaks down proteins to amino-acids. The trypsin is only formed when trypsinogen is mixed with enterokinase which is secreted by the walls of the duodenum. The pancreas also produces the hormone insulin which regulates the sugar content of the blood.

No food is absorbed in the mouth, stomach or upper part of the small intestine, but absorption begins in the next part of the small intestine, the ileum.

Reproduction

Q. 26:

Give an account of (*a*) sexual reproduction and (*b*) vegetative reproduction in a NAMED flowering plant. In what ways would you expect the offspring produced by vegetative reproduction to differ from those produced by sexual reproduction?

A: (*a*) Sexual Reproduction in the Creeping Buttercup.

Sexual reproduction involves the fusion of two gamete nuclei, which leads to the doubling of the number of chromosomes in the nucleus of the zygote so formed. The zygote will ultimately grow into a new individual.

The flower of the Creeping Buttercup has five green, separate sepals, five yellow, separate petals which have honey guides on the upper surface, and a nectary covered by a scale at its inner side. There are numerous stamens, arranged in a whorl outside a similar whorl of numerous carpels.

Pollination is brought about by insects which are attracted by the colour of the petals, and in their search for nectar they transfer pollen from the anthers to the stigmas of the carpels in the same or another flower.

The pollen germinates on the surface of the stigma, and the pollen tube grows down through the style, ultimately reaching the ovule which it penetrates. One of the two nuclei in the pollen tube fuses with the egg nucleus and so initiates the growth of a new individual. Fusion of the second nucleus in the pollen tube with other nuclei in the ovule starts a rapid division of these nuclei resulting in the formation of the endosperm. The zygote divides so that the radicle, plumule, and cotyledons are differentiated. During this time the ovary wall enlarges and ultimately dries out to form a single-seeded, non-dehiscent achene which on germination of the enclosed seed will produce a new individual.

(*b*) Vegetative Reproduction in the Creeping Buttercup

The Creeping Buttercup reproduces asexually by overground runners, or *stolons*. The stem of the plant is very short, so that the leaf bases are near the ground. The buds in the axils of the leaves elongate rapidly to produce long stems — the stolons — which run along the ground. Where the nodes touch the ground, adventitious roots develop under suitable conditions of soil moisture, and the bud in the axil of the leaf of the node produces a new stem which develops into a new plant. This procedure is repeated, so that several daughter plants are formed on the same runner.

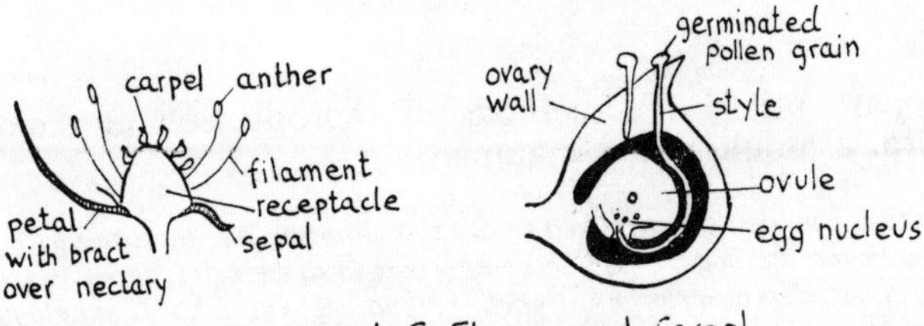

Creeping Buttercup: L.S. Flower and Carpel

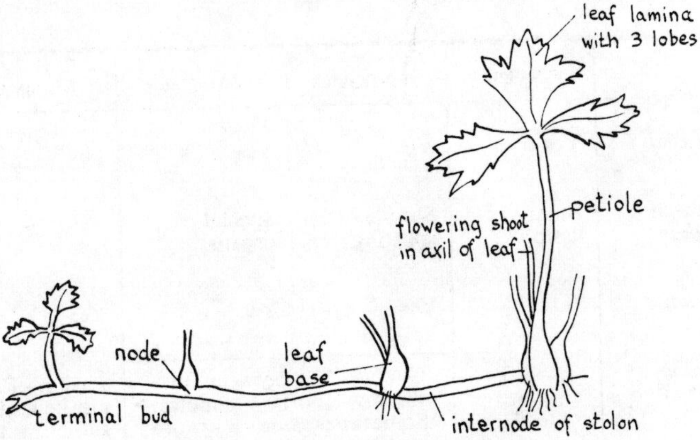

Vegetative Reproduction: Creeping Buttercup.

Offspring produced sexually or asexually will show essential similarities to the parents, in that they will all be recognisable as Creeping Buttercups; but, as in vegetative reproduction there has been no fusion of gametes and consequently no recombination of genes, the offspring should be identical with the parents except for the minor differences induced by environmental conditions.

Where sexual reproduction has taken place, there has been a recombination of genes, so that the offspring can vary in both external features (size of leaf, density of flower colour, *etc.*), and physiological features (ability to resist drought, greater fertility, *etc.*).

[*Do not confuse pollination and fertilization. Pollination is the transfer of pollen to the stigma, and fertilization is the fusion of the two gamete nuclei.*]

Q. 27:

Compare briefly the methods of reproduction in Amoeba, Spirogyra, a mould and a mammal.

[*This question can well be answered by a table pointing out the similarities and differences in the methods. As the comparison is to be brief, there is no need to go into a detailed description of each method. If the table is supplemented by well-labelled diagrams, this should be sufficient.*]

A:

	AMOEBA	SPIROGYRA	MUCOR	MAMMAL
Asexual	Yes	Yes	Yes	No
Asexual method	Binary fission	Simple splitting	Special organs	—
Sexual	No	Yes	Yes	Yes
Sexual method	—	Conjugation between two slightly motile gametes. Sometimes one is more motile than the other.		Fusion of motile male gamete with static female gamete.
Asexual encystment	Yes	No	No	No
Protection of zygote	—	Cyst	Cyst	Protected and fed in female body.
Product from cyst	Numerous new individuals	Single filament	Sporangium	—
Reduction division	—	On germination of the cyst	—	At gamete formation

The formation of a cyst is usually a response to adverse environmental conditions.

Stages in Binary Fission in Amoeba

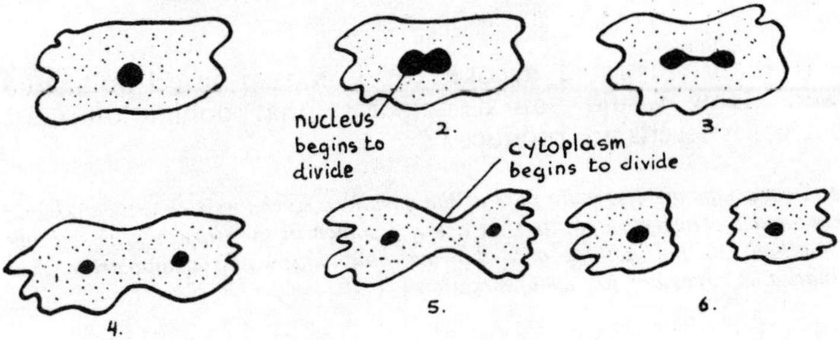

Reproduction of Mucor

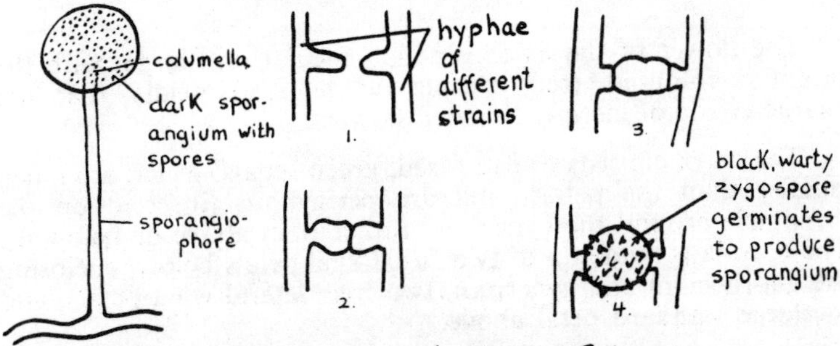

Sexual Reproduction in Spirogyra

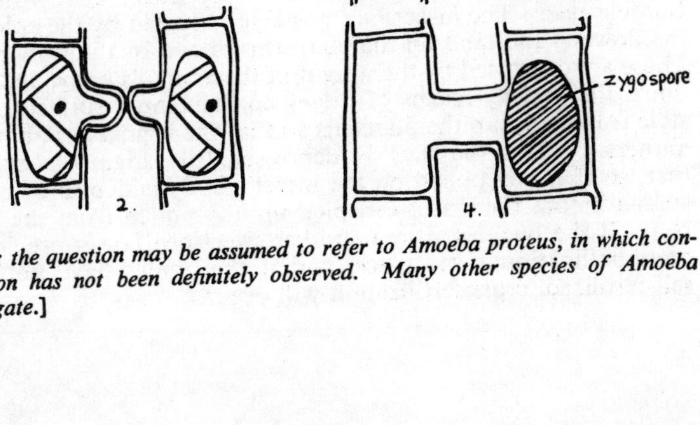

[*Note: the question may be assumed to refer to Amoeba proteus, in which conjugation has not been definitely observed. Many other species of Amoeba conjugate.*]

Q. 28:

Give an illustrated account of a NAMED insect-pollinated flower. How would you demonstrate that pollination must occur before seeds are produced?

[*The illustrations are the main part of the question, so the written answer should be confined to clarifying points that are not shown in the drawings, describing the methods of pollination, and showing how you would demonstrate that pollination is necessary for seed formation.*]

A: The Sweet Pea

The flower of the Sweet Pea is irregular and large, with the brightly coloured (red, purple, white *etc.*) petals that are characteristic of insect-pollinated flowers.

The calyx consists of five fused, green sepals, which are much smaller than the petals, and are perigynous (inserted on the receptacle around the gynecium, rather than above or below it). The corolla is made up of two fused keel petals below, enclosing the androecium and gynecium, two free, lateral wing petals, and the large standard petal above.

There are ten stamens; the filaments of the nine lower ones are fused, enclosing the nectaries, and the upper one is free. The anthers are free and introse. The gynecium is a single superior carpel enclosing several ovules. After fertilization, the carpel matures to form a legume.

The pollen and ovules are mature before the flower opens, so that self-pollination often occurs before there is a chance for cross-pollination to be brought about by heavy insects like the Bumble Bee. The insects are possibly attracted by the colour of the flowers, and land on the platform formed by the wing petals. These are depressed by the weight of the animal's body, and press down the keel by means of ridges on their inner surfaces. The style is longer than the filaments so that the stigma lies above the anthers. When the keel is depressed, the stigma is exposed first, so that any pollen on the insect's body is deposited on the stigma before the insect can pick up any pollen from the sweet pea. If the Bee has visited another sweet pea flower previously, cross-pollination takes place, and if the ovules have not been self-fertilized, cross-fertilization will occur.

THE SWEET PEA

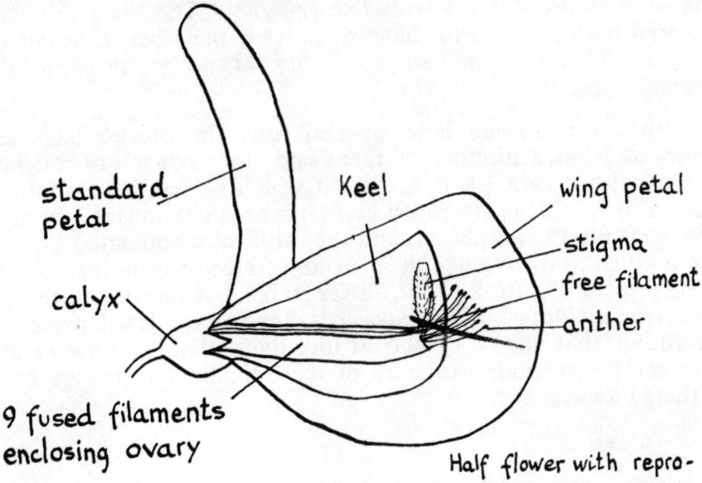

Half flower with reproductive parts intact.

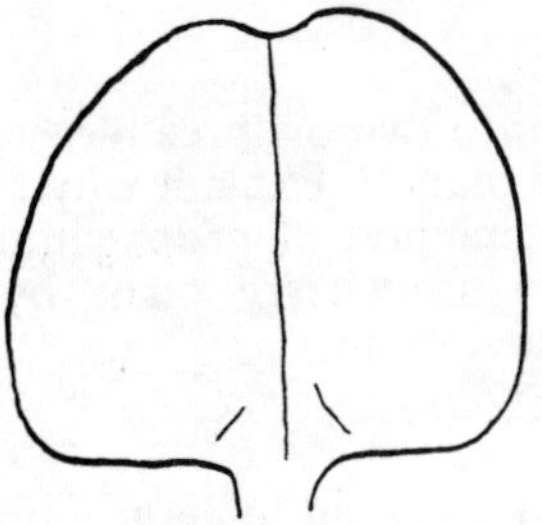

STANDARD PETAL

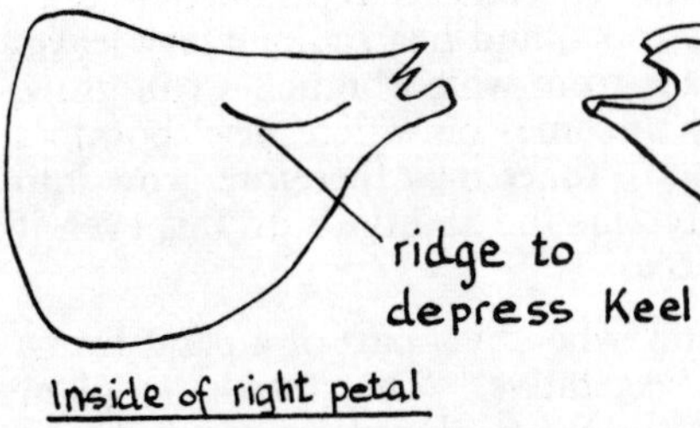

Inside of right petal

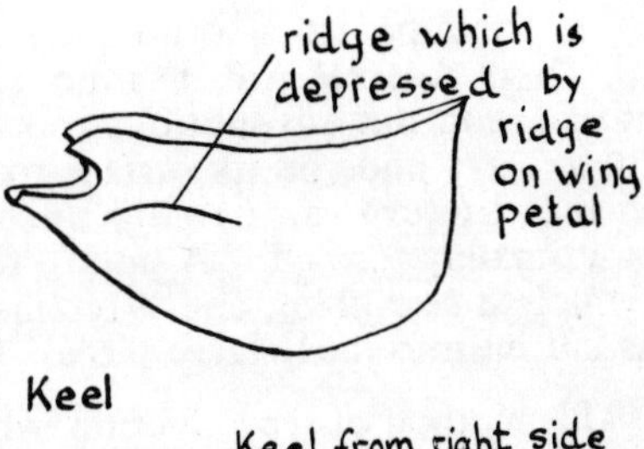

Keel from right side

To demonstrate that pollination is necessary for the formation of seeds, a sweet pea plant with several flower buds is selected. The young buds are opened gently before the pollen is ripe, and the anthers are cut off with fine scissors. The buds are then covered with fine muslin bags to prevent their being visited by insects. This removes all possibility of pollen reaching the stigma.

When the flowers have opened fully the muslin bags are removed from a number of them and the stigmas are touched lightly with a camel-hair brush on which pollen from another, normally-flowering, sweet pea plant has been previously collected. The bags are then replaced, and the artificially pollinated flowers are marked to distinguish them by any convenient means, *e.g.* by tying a piece of cotton to the flower stalks. A few days after the petals have dropped off the bags are taken away. It will presently be found that many of the artificially pollinated flowers are producing seed pods, while all of the unpollinated flowers have withered away.

Q. 29:

If a 'seed' potato is cut into two or three pieces, each piece, if planted will produce a new plant. Explain why this is possible. Compare and contrast this method of propagation with that of (*a*) the multiplication of a strawberry plant by runners and (*b*) sowing seeds of an annual plant.

A: A 'seed' potato is a tuber, *i.e.* an underground stem swollen with food reserves. It has no specialized reproductive cells, but can send out adventitious roots, and has rudimentary leaves with axillary buds on its surface from which branches can grow, and these in turn can develop rhizomes on which 'new' potatoes are ultimately formed. A potato tuber may therefore grow into a complete new plant, and it retains the ability to do this even if it is cut up into fairly large pieces.

This method of reproduction whereby a part of a plant forms an entire new plant, is called vegetative. The strawberry plant provides another example of it. When flowering comes to an end, the buds in the axils of the leaves grow into thin, long, horizontal stems with a few scale leaves. These runners are not

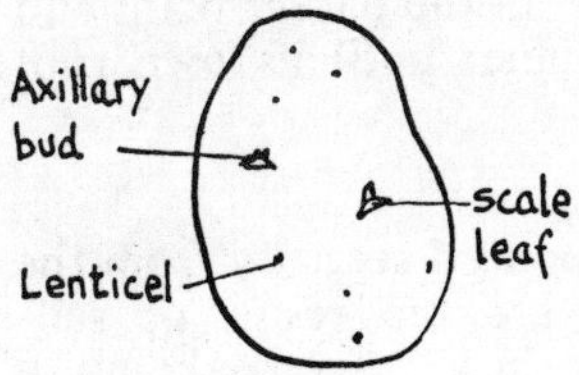

New Potato

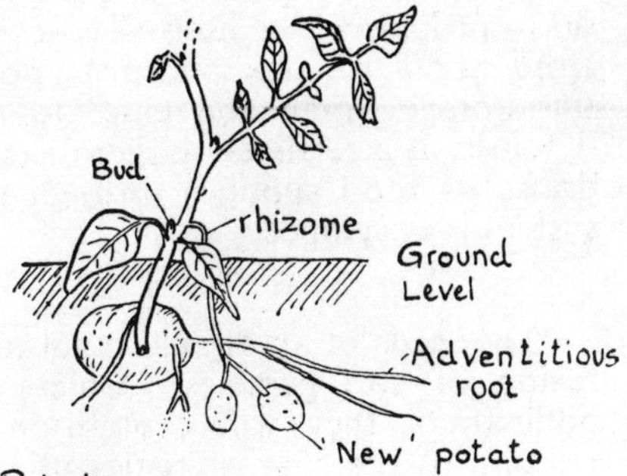

Potato:
Vegetative Reproduction

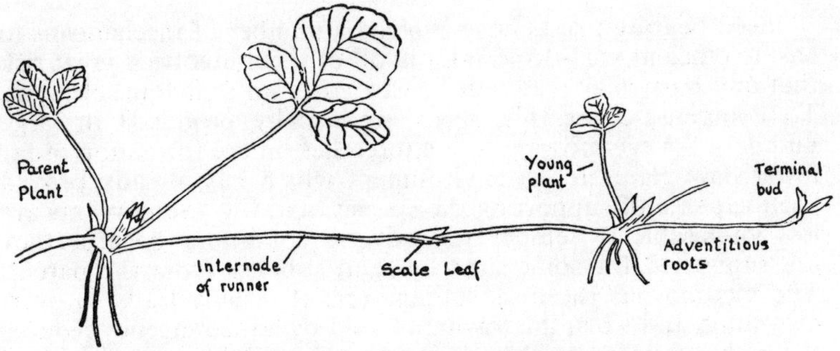

Strawberry: Vegetative Reproduction.

rhizomes because they lie on the surface of the ground, but each terminal bud can grow upwards into a new plant and can grow adventitious roots, and the new plant can send out further runners in its turn. When the runner between an old and a new plant rots away, the latter becomes an independent individual.

Thus both the potato and the strawberry can — and normally do — propagate themselves vegetatively and therefore asexually from modified stems, but in the potato the new plant grows from the storage organ and close to it, while in the strawberry it develops at a relatively larger distance from the parent plant and draws its food supplies through the runner until its own root system is developed.

The seeds of an annual plant are produced sexually from the fusion of two gametes. Unless they are the result of self-pollination, they will contain in their hereditary make-up a mixture of the characteristics of two parent plants of the same species. This improves the chances of evolutionary progress of the species, compared with vegetatively reproduced individuals which must in all essential respects be identical to the single parent plant from which they stem.

Seed-bearing plants have evolved a number of mechanisms to ensure efficient seed-dispersal, but all of them involve a great risk that any particular seed will be carried to an unfavourable spot. To compensate for this, seeds are usually produced in large numbers. Vegetative reproduction relies on the formation of far fewer new plants in an environment which has already proved itself capable of supporting the species, and the young plants are less vulnerable to temporarily adverse conditions because they are supported for some time by food supplies from the parent. The method has the disadvantage that it usually leads to over-crowding, and when the sowing of seed by human agency reduces the risk of unfavourable dispersal to a minimum, the sexual method has a clear advantage for those species which can successfully ripen seeds in a given climate and endow them with a reasonable supply of nutrients. This is not particularly true of potatoes and strawberries, and for this reason their cultivation generally makes use of the vegetative method. Moreover, when they are grown commercially, it becomes important to maintain the characteristics of a variety, and if this variety is itself a hybrid, multiplication by the sexual method would involve cross-fertilization, and the next generation of seed would contain diverging types.

[*The comparison and contrasts should be between the methods rather than the species.*]

Q. 30:

Make a large, clearly labelled drawing of the reproductive organs of the female rabbit (or other NAMED mammal). Explain clearly (*a*) the functions of the placenta and (*b*) how the young mammal is cared for by its parents both before and after birth.

A:

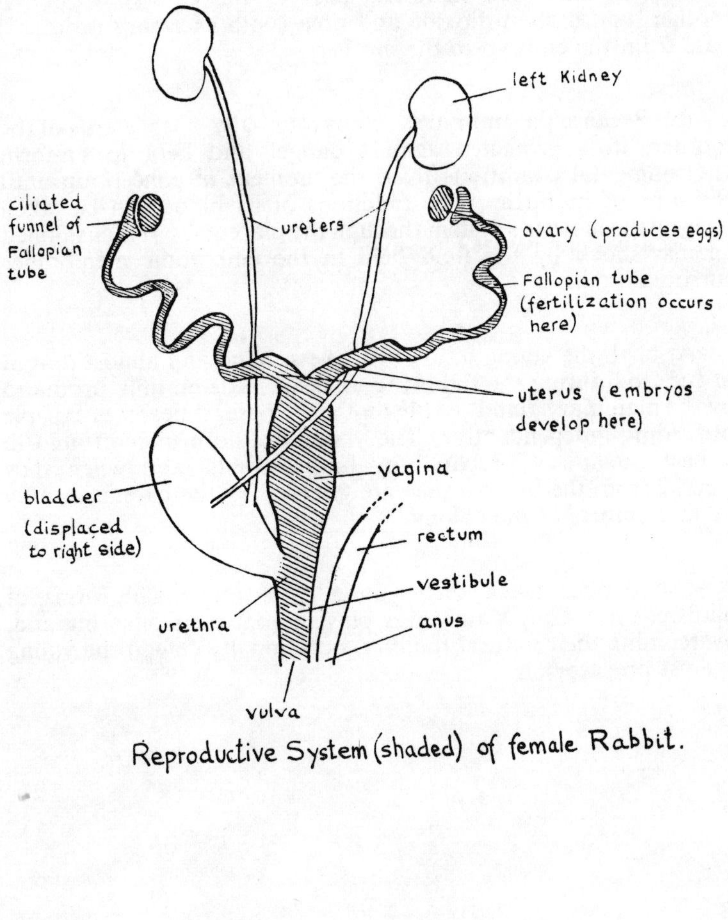

Reproductive System (shaded) of female Rabbit.

(*a*) The placenta is made up of embryonic and maternal tissues in close connection. The blood system of the embryo develops early and runs into the placenta through the umbilical cord. The membranes separating the blood vessels of the uterus from those of the embryo are very thin, allowing small molecules to pass from the mother's blood to the embryo, and from the embryo to its mother.

Oxygen and food materials pass to the embryo from the mother, and carbon dioxide and urea (both excretory products) pass from the embryo to the mother.

(*b*) Because the embryo develops internally in the uterus of the mother, it is protected against danger and kept in uniform environmental conditions from the moment of conception until the time of its birth. The functions of its blood can be taken over by the mother's system through the placenta. It is cushioned against shock by the fluid held in the embryonic membranes surrounding it.

At birth the young rabbit is helpless, blind and almost devoid of fur, and during the following weeks it feeds on milk produced by the mammary glands of the mother. Later it becomes capable of feeding independently. The young are born underground in a nest previously prepared by both parents, and when they emerge from the burrow they are guarded by the parents and by other members of the colony.

The parents teach their young by example and forms of punishment. They watch over play, groom their offspring and, overcoming their natural timidity, occasionally defend the young against predators.

Irritability

Q. 31:

Living organisms are said to exhibit irritability. Explain the meaning of this statement by reference to the behaviour of (*a*) the shoot and root of a germinating seedling and (*b*) the adult stage of an insect.

A: Irritability is the capacity to respond to external and internal stimuli. It is one of the characteristics which help to distinguish living organisms from inanimate matter. As a general rule, irritability in living protoplasm takes the form of changes or movements which derive their energy from within the living tissue, and whose magnitude is independent of the strength of the stimulus; the reactions of dead matter to outside influences usually derive their energy from without and vary in proportion to the strength of the influence.

A germinating seedling exhibits irritability in a number of ways, of which the most obvious ones are the tropisms — growth in a direction which is determined by the direction of a stimulus. The root of a seedling is positively, and its stem negatively, geotropic in response to the stimulus of gravity. The growing tip of a seedling shows positive phototropism by bending towards the source of maximum light. Root hairs grow towards regions of optimum moisture (hydrotropism), and pea tendrils respond to contact by twining. With the exception of the last example, these reactions are all tropisms, and although they do depend somewhat on the strength of the stimulus their energy is supplied entirely by the plant. In many cases, the external stimulus acts on the seedling by affecting the production and flow of the growth-promoting auxins and thereby setting up internal chemical stimuli. In this way, one part of the seedling may respond to a stimulus detected by another part. This is an example of conductivity, which is an important component of irritability. That the organism as a whole may be affected by very localized reactions to stimuli is illustrated by the action of the guard cells on green leaves, which by a chemical response to light and darkness open and close the stomata and so control the transpiration of the entire plant.

In an adult insect, irritability takes quite different forms because conductivity is greatly increased by the presence of a nervous system. An adult insect can perceive light, heat, shapes, movement of external objects, odours, vibrations, and the physical properties of surfaces instantly by means of its sophisticated sensory apparatus. Any one of these may excite irritability, and the response takes the form of movement rather than growth; *e.g.* a moth may be attracted by bright light, and a

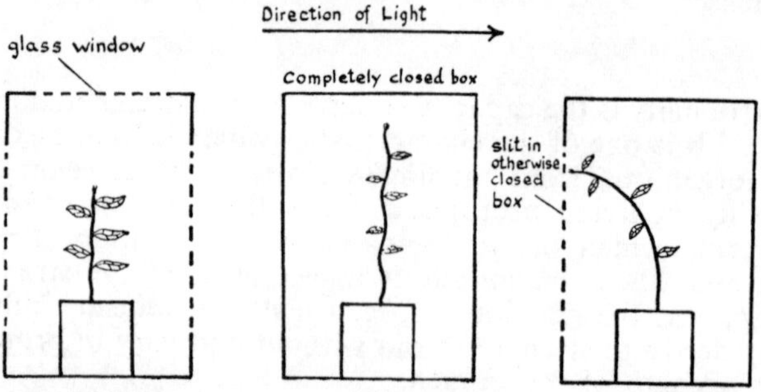

fly may escape from a sudden movement near it. Several simultaneous stimuli may tend to evoke conflicting responses, *e.g.* a honey bee may be attracted by the odour of nectar but repelled by the odour of other foraging bees from a different hive. In such a case, the central nervous system may evaluate the importance of the different stimuli and decide which response is the most appropriate — not always correctly. Internal stimuli such as hunger or impulses derived from the release of hormones at a particular stage of the reproductive cycle add to the almost incessant stream of outside influences. The behaviour of an adult insect is therefore far more complicated than that of a plant, or even that of protozoa and primitive metazoa, which respond chiefly by simply moving towards or away from an external stimulus (taxis) — so much so, that one is tempted to wonder whether the insect exhibits yet another aspect of irritability which we recognise in ourselves as the will.

[*The question must be answered by reference to a germinating seedling and an adult insect. This means that the examples quoted must not be drawn from other types of organism, but the seedling or insect need not be named, and details of experimental demonstrations are not required. The answer may be written as a general essay, but each sentence should contribute some valid and relevant point.*]

Circulation

Q. 32:

Explain how the circulation of the blood of a mammal causes a red blood cell to go through the lungs between every alternate passage through the heart.

[*This question needs careful interpretation. The key word is 'causes', which means that a description of how the blood is circulated is necessary. Secondly, the passage through the lungs is particularly emphasized, but to show the alternate passage through the heart, the main parts of the rest of the circulatory system have to be mentioned.*]

A: The mammal has a double circulation, which means that a particular red blood cell passes through the heart twice in a complete circulation, and that the right and left sides of the heart are separate. During one half of this cycle the arterial blood supplies the body tissues with oxygen and returns to the right auricle of the heart as venous blood; during the second half the venous blood is re-oxygenated and returned to the left auricle. The details are as follows:

The blood leaves the left ventricle via the aortic arch which branches to all parts of the body, where the very fine branches of the subsidiary arteries form capillaries. The capillaries join to form veins which empty into the venae cavae, which, in turn, open into the right auricle. The blood then passes through the tricuspid valve into the right ventricle whence it is pumped into the lungs through the pulmonary artery. (The blood has passed through the heart once.) The pulmonary artery branches to form capillaries in the lungs, and these rejoin to form the pulmonary vein. The blood passes through the pulmonary vein into the left auricle, and on through the mitral valve and the left ventricle into the aorta. (The blood has now passed through the heart twice to complete the circuit.) Thus each blood cell passes through the lungs and body alternately.

This circulation is maintained by the action of the heart. In this particular case we are interested in the right side of the heart, as it is expressly concerned with the passage of blood through the lungs. The blood enters the right auricle through the venae cavae during diastole (relaxed period) of the heart, which occurs between every heart beat. During systole, a wave of contraction passes over the muscles of the heart, affecting first the auricle, and then the ventricle, so that the blood is forced from the auricle into the ventricle through the tricuspid valve; and as the contraction reaches the ventricle, the blood is forced into the relaxed pulmonary artery, hence to the lungs, and through them. The tricuspid valve is prevented from inverting when the ventricle contracts by the chordae tendinae. The opening of the pulmonary artery has three semi-lunar valves which prevent the blood from refluxing into the ventricle, and additional valves in the veins guard against a local reversal of flow.

Multiple Choice Questions

In a Multiple Choice Test you are required to select the best answer to each question from several (usually five) suggested answers. Only one answer will be the best in each case: usually, but not always, the others will be wrong.

The ability to answer such questions will develop naturally as you work through your normal course. Multiple Choice questions are not in themselves suitable for preparation purposes, but are useful for tests and self-assessment. This Section can do no more than give a few illustrative examples of the types of question you may expect. The answers are given on p. 80.

Read the questions very carefully, because often a single word determines your answer. The very fact that only one answer is wholly correct can give you a valuable clue. All questions carry equal marks.

Questions 33–36, Directions : Each of these questions or incomplete statements is followed by five suggested answers. Select the best answer in each case and mark the answer sheet accordingly.

Q. 33 : The body of *Hydra* (or Sea Anemone) has two layers of cells known as

A protoplasm and ectoplasm
B epidermis and epithelium
C epithelium and protoplasm
D ectoplasm and endoplasm
E ectoderm and endoderm

Q. 34 : In mammals the embryo develops in the

A ovary **B** placenta
C uterus **D** ureter
E urethra

Q. 35 : In a parasitic relationship between two organisms

A both members benefit **B** neither member is harmed
C both members suffer harm **D** only one member benefits
E neither member gains benefit

Q. 36 : Which of the following is NOT a characteristic feature of a carnivorous mammal?

A The presence of canine teeth, well developed for defence and attack
B The presence of incisor teeth, which have open roots and grow continuously through life
C Limited side-to-side jaw movements
D The presence of carnassial teeth, well designed for cutting flesh
E The presence of molar teeth near the back of the jaw

Questions 37–39 relate to the adjacent diagram of a longitudinal section through a flower.

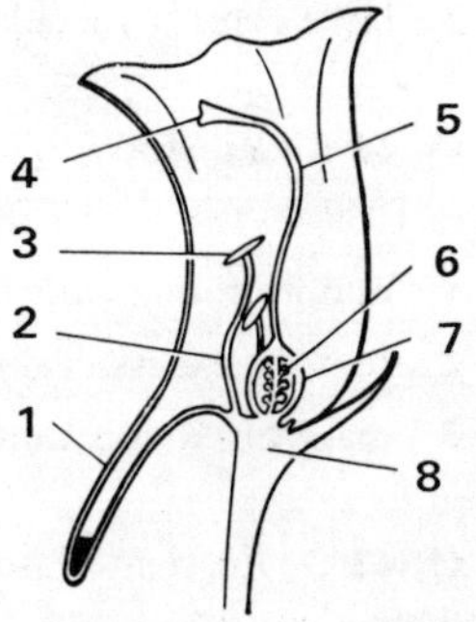

Q. 37 : Fertilization takes place in

A 1 **B** 2 **C** 3 **D** 6 **E** 8

Q. 38 : The male gametes are formed in

A 2 **B** 3 **C** 4 **D** 5 **E** 7

Q. 39 : This flower is normally cross-pollinated. Which of the following would best ensure that self-pollination did NOT occur?

A There are honey guide lines on the petals

B The flower is not regular in shape

C The anthers are large

D The ovary contains many ovules

E The ripe stigma overarches the stamens

Questions 40–42, Directions : For each of the questions below, ONE or MORE of the responses are correct. Decide which of the responses is (are) correct. Then choose :

A if **1, 2** and **3** are all correct

B if **1** and **2** only are correct

C if **2** and **3** only are correct

D if **1** only is correct

E if **3** only is correct

Q. 40 : Roots of flowering plants

1 do not respire

2 contain chlorophyll

3 have central vascular tissue

Q. 41 : In studying the rate at which small mammals consume oxygen it would be necessary to

1 maintain constant temperature

2 remove excess carbon dioxide

3 maintain a constant supply of oxygen

Q. 42 : Efficient gaseous exchange in the lungs of a mammal is dependent upon

1 a large surface area

2 a respiratory pigment, haemoglobin

3 movement of the ribs and diaphragm

Questions 43–47, Directions : Each of the following questions consists of a statement in the left-hand column followed by a second statement in the right-hand column. Decide whether the **first** statement is true or false. Decide whether the **second** statement is true or false. Then, on the answer sheet, mark :

A if both statements are true and the second statement is **a correct explanation** of the first statement

B if both statements are true and the second statement is **NOT a correct explanation** of the first statement

C if the first statement is true but the second statement is false

D if the first statement is false but the second statement is true

E if both statements are false

	First Statement	Second Statement
Q. 43 :	Green plants form the first link in a food chain	Green plants are able to synthesize food by means of light energy
Q. 44 :	Any part of the shoot or root of flowering plants can contribute to growth in length	All the cells of flowering plants are unspecialized
Q. 45 :	Protozoa have no specialized respiratory surfaces	Many protozoa possess contractile vacuoles
Q. 46 :	In mammals the lymphatic vessels may at times appear milky	In mammals the lymph becomes laden with tiny droplets of fat after a meal rich in fats
Q. 47 :	Green plants turn yellow when grown in the dark	Sunlight is necessary for the production of sugar by photosynthesis

Questions 48–50, Directions : Each group of questions below consists of five lettered headings followed by a list of numbered questions. For each numbered question select the one heading which is most closely related to it. Each heading may be used once, more than once or not at all. The questions relate to the following structures of a bony fish:

A Scales

B Gills with a good blood supply

C Powerful muscular tail

D Paired pectoral and pelvic fins

E Unpaired dorsal and ventral fins

Relate each of the descriptions below with one of the lettered headings above.

Q. 48 : Keeps fish on an even keel and prevents rolling in the water

Q. 49 : Forms respiratory surface of fish

Q. 50 : Enables fish to alter position in depth in the water

Questions 51–54 refer to the adjacent diagram of an experiment to demonstrate that carbon dioxide is essential for photosynthesis.

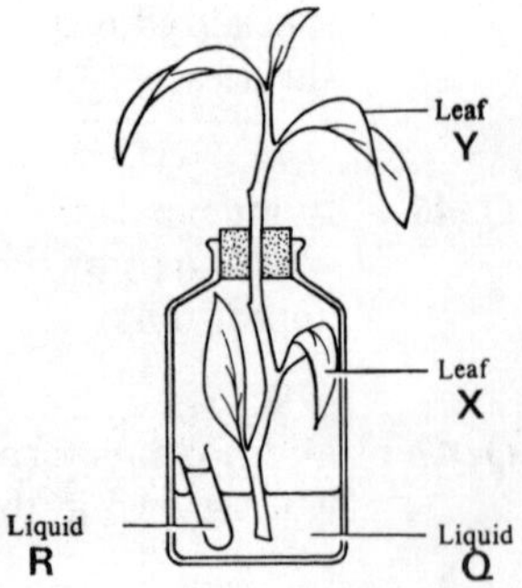

Q. 51 : Before setting up the experiment, the plant shoot used should be

A weighed

B boiled in methylated spirits

C boiled in water

D kept in the light for several hours

E kept in the dark for several hours

Q. 52 : The liquid R is a solution or suspension of

A lime

B cobalt chloride

C calcium chloride

D alkaline pyrogallol

E sodium or potassium hydroxide

Q. 53 : The purpose of the liquid R is to

A absorb oxygen

B absorb carbon dioxide

C supply oxygen

D supply carbon dioxide

E dry the air in the jar

Q. 54 : After the experiment had been set up for eight hours of daylight, which of the following is likely to be true?

A Both leaves (X andY) contain starch

B Leaf X contains starch; leaf Y does not contain starch

C Leaf X does not contain starch; leaf Y does contain starch

D Leaf X is green; leaf Y is yellow

E Leaf X is yellow; leaf Y is green

STOP

Now go back and check your work.

MODEL TEST PAPER INDEX

The following groups of questions are each equivalent to a typical Paper 2 of the London University G.C.E.
Four questions must be chosen from each paper.
All questions carry equal marks.

Time: 2 hours.

Paper I: QQ. 1, 5, 8, 13, 15, 25, 26, 32.

Paper II: QQ. 2, 3, 9, 11, 18, 22, 23, 30.

Paper III: QQ. 4, 7, 14, 20, 21, 24, 29, 31.

PAPER 1 ($1\frac{1}{4}$ hours) consists of 60 multiple choice questions to be answered on the question-paper, all of which are to be attempted. The paper may include simple questions involving elementary identification, comments on features of biological interest, observations on simple experiments and interpretation of experimental data. Some representative examples are given in the section beginning on p. 74.

PAPER 1 carries four-tenths of the total subject marks, and Paper 2 six-tenths.

Answers to Multiple Choice Questions

Q. 33 : E	Q. 41 : A	Q. 49 : B
Q. 34 : C	Q. 42 : A	Q. 50 : D
Q. 35 : D	Q. 43 : A	Q. 51 : E
Q. 36 : B	Q. 44 : E	Q. 52 : E
Q. 37 : D	Q. 45 : B	Q. 53 : B
Q. 38 : B	Q. 46 : A	Q. 54 : C
Q. 39 : E	Q. 47 : B	
Q. 40 : E	Q. 48 : E	